Inverse Heat Conduction and Heat Exchangers

Edited by Suvanjan Bhattacharya,
Mohammad Moghimi Ardekani,
Ranjib Biswas and R. C. Mehta

Published in London, United Kingdom

IntechOpen

Supporting open minds since 2005

Inverse Heat Conduction and Heat Exchangers
http://dx.doi.org/10.5772/intechopen.80096
Edited by Suvanjan Bhattacharya, Mohammad Moghimi Ardekani, Ranjib Biswas and R. C. Mehta

Contributors
Animesh Talapatra, Debasis Datta, Rania Jradi, Ali Fguiri, Christophe Marvillet, Mohamed Razak Jeday, Sergio García, Alfredo Trueba, Eman Mansour, Tao Min, Bin Chen, Jia-Meng Tian, Zhifu Zhou, Suvanjan Bhattacharya, Devendra K Vishwakarma, Sanghati Roy, Ranjib Biswas, Mohammad Moghimi Ardekani, R. C. Mehta, Hussain H. H. Al-Kayiem, Muna S. Kassim, Saud Tameem Taher, Sergey Voronin, Xing Chen, Yao Sun, Qiang Huang

First published in London, United Kingdom, 2020 by IntechOpen
IntechOpen is the global imprint of INTECHOPEN LIMITED, registered in England and Wales, registration number: 11086078, 5 Princes Gate Court, London, SW7 2QJ, United Kingdom
Printed in Croatia

British Library Cataloguing-in-Publication Data
A catalogue record for this book is available from the British Library

Additional hard and PDF copies can be obtained from orders@intechopen.com

Inverse Heat Conduction and Heat Exchangers
Edited by Suvanjan Bhattacharya, Mohammad Moghimi Ardekani, Ranjib Biswas and R. C. Mehta
p. cm.
Print ISBN 978-1-78985-177-9
Online ISBN 978-1-78985-178-6
eBook (PDF) ISBN 978-1-83962-384-4

For EU product safety concerns:
IN TECH d.o.o., Prolaz Marije Krucifikse Kozulić 3, 51000 Rijeka, Croatia, info@intechopen.com or visit our website at intechopen.com.

Meet the editors

Dr. Suvanjan Bhattacharyya is currently working as an Assistant Professor in the Department of Mechanical Engineering of Birla Institute of Technology & Science, Pilani, Pilani Campus, Rajasthan, India. Dr. Bhattacharyya completed his post-doctoral research from the University of Pretoria, South Africa, under the supervision of Prof. Josua P Meyer. Dr. Bhattacharyya completed his PhD in Mechanical Engineering from Jadavpur University, Kolkata, India and with the collaboration of Dusseldorf University of Applied Sciences, Germany. He received his Master's degree from the Indian Institute of Engineering, Science and Technology, India (formerly known as Bengal Engineering and Science University), on Heat-Power Engineering. His research interests are in computational fluid dynamics in fluid flow and heat transfer, specializing in laminar, turbulent, steady, unsteady separated flows and convective heat transfer, experimental heat transfer enhancement, solar energy and renewable energy. He is the author and co-author of 85 papers in high ranked journals and prestigious conference proceedings. He has been awarded the best paper award in a number of international conferences. He is also on the editorial boards of 11 journals and is a reviewer of more than 25 prestigious journals.

Currently, Dr. Moghimi Ardekani is working in the Department of Mechanical and Aeronautical Engineering at the University of Pretoria, South Africa. His professional experience includes several roles in academics and industries since 2007. He has more than three years of industrial experience in design of heat exchangers and pressure vessels for the oil, gas, and petrochemical industries. In addition, he is also interested in solar thermal plants. He is the author and co-author of 42 papers in high ranked journals and prestigious conference proceedings as well as a book on design of heat exchangers. He is on the editorial boards of 4 international journals and a reviewer of 10 ISI journals. In addition, his research findings gained him the international prize of Green Talent from the German Ministry of Research and Education in 2015. He has done several collaborations and consultations with different research institutes and companies around the world. To name a few: DLR (Germany), SOLASTOR (Australia), University of Nottingham (UK), University of Clarkson (USA), Concordia university (Canada).

Dr. Ranjib Biswas completed his B.Tech. in Mechanical Engineering from Kalyani Govt. Engineering College in 1999. Later on he did his Masters and Ph.D. (Engineering) from Jadavpur University. His area of interest includes additively manufactured, 3D printed heat exchangers, layer-by-layer line welding, process intensification, laser beam machining, micromachining etc. He has published several research papers in international journals of repute. Dr. Biswas joined the MCKV Institute of Engineering (MCKVIE) in 2003 as Lecturer and is presently working as Associate Professor of the Mechanical

Engineering Department. He is deeply involved in the assessment and accreditation (A&A) process of different programs of the institute as Coordinator of Internal Quality Assurance Cell (IQAC) of MCKVIE. Dr. Biswas has also organized several national and internal conferences, workshops, seminars, FDPs etc.

R. C. Mehta obtained his Ph.D. from the Indian Institute of Technology, Madras. He has worked as the Head of Aerodynamics Division of Vikram Sarabhai Space Centre/Indian Space Research Organization. He has participated in the design of launch and reentry vehicles. He has served as a Senior Fellow in the School of Mechanical and Aerospace Engineering at Nanyang Technological University, Singapore. He is the recipient of the Life Time Achievement Award from the Flow Physics Society of India. He is a senior member of AIAA. He has published over 120 papers in peer-reviewed national and international journals. He has published five chapters and co-authored two books. He is a reviewer for many international journals. He is presently Dean in the Noorul Islam Centre for Higher Education, Kumaracoil, India.

Contents

Preface XI

Section 1
Heat Exchanger 1

Chapter 1 3
Applications of Heat Transfer Enhancement Techniques: A State-of-the-Art Review
by Suvanjan Bhattacharyya, Devendra K. Vishwakarma, Sanghati Roy, Ranjib Biswas and Mohammad Moghimi Ardekani

Chapter 2 23
Applications of Compound Nanotechnology and Twisted Inserts for Enhanced Heat Transfer
by Hussain H. Al-Kayiem, Muna S. Kassim and Saud T. Taher

Chapter 3 47
Tubular Heat Exchanger Fouling in Phosphoric Acid Concentration Process
by Rania Jradi, Ali Fguiri, Christophe Marvillet and Mohamed Razak Jeday

Chapter 4 63
Fouling in Heat Exchangers
by Sergio García and Alfredo Trueba

Chapter 5 89
Molecular Dynamics Simulation-Based Study on Enhancing Thermal Properties of Graphene-Reinforced Thermoplastic Polyurethane Nanocomposite for Heat Exchanger Materials
by Animesh Talapatra and Debasis Datta

Section 2
Inverse Problem 103

Chapter 6 105
A Numerical Approach to Solving an Inverse Heat Conduction Problem Using the Levenberg-Marquardt Algorithm
by Tao Min, Xing Chen, Yao Sun and Qiang Huang

Chapter 7 125
The Fundamental and Application of Surface Heat Flux Estimation by Inverse Method in Cryogen Spray Cooling
by Bin Chen, Jia-Meng Tian and Zhi-Fu Zhou

Chapter 8 145
Equation of State
by Eman Mohamed Mansour

Chapter 9 163
Influence of Input Parameters on the Solution of Inverse Heat Conduction Problem
by Rakhab C. Mehta

Chapter 10 179
Introduction to Numerical Approaches for Forward and Inverse Heat Transfer Problems
by Sergey Voronin

Preface

This edited volume is a collection of reviewed and relevant research chapters concerning the developments within the inverse heat conduction and heat exchangers field of study. The book includes scholarly contributions by various authors and it has been edited by a group of experts pertinent to heat conduction and heat exchangers. Each contribution comes as a separate chapter complete in itself but directly related to the book's topics and objectives.

The book is divided in two sections: Heat Exchanger and Inverse Problem. The Heat Exchanger section includes chapters dealing with the topics: "Application of Heat Transfer Enhancement Techniques: A State-of-the-Art Review", Applications of Compound Nanotechnology and Twisted Inserts for Enhanced Heat Transfer", "Tubular Heat Exchanger Fouling in Phosphoric Acid Concentration Process", "Fouling in Heat Exchanger and Molecular Dynamics" "Simulations Based Study on Enhancing Thermal Properties of Graphene Reinforced Thermoplastic Polyurethane Nanocomposite for Heat Exchanger Materials". The following section, Inverse Problem, includes "A Numerical Approach to Solving an Inverse Heat Conduction Problem Using the Levenberg-Marquardt Algorithm", "The Fundamental and Application of Surface Heat Flux Estimation by Inverse Method in Spray Cooling", "Equation of State (EOS)", "Influence of Input Parameters on the Solution of Inverse Heat Conduction Problem", "Introduction to Numerical Approaches for Forward and Inverse Heat Transfer Problems".

The target audience comprises scholars and specialists in the field.

InTechOpen

Section 1

Heat Exchanger

Chapter 1

Applications of Heat Transfer Enhancement Techniques: A State-of-the-Art Review

Suvanjan Bhattacharyya, Devendra K. Vishwakarma, Sanghati Roy, Ranjib Biswas and Mohammad Moghimi Ardekani

Abstract

The fundamentals of heat transfer and its applications, the classification of heat transfer technology and different heat transfer techniques, and the needs for augmentation and its benefits and the different combinations of two or more inserts and integral roughness elements for heat transfer augmentation purpose have been introduced and discussed in this chapter. It is shown that most of the compound techniques performed better than the individual inserts for heat transfer enhancement. This chapter has also been dedicated to understanding the basic concepts of vortex generators for heat transfer enhancement in plate-fin heat exchangers. The performance of transverse, longitudinal, and wing-type vortex generators has been discussed as well.

Keywords: heat transfer, review, enhancement, heat exchanger, vortex generators, twisted tape, ribs, combine techniques

1. Introduction

The phenomenon of heat transfer has always been a topic of interest to researchers and manufacturers alike. The previous researchers have addressed heat transfer characteristics of wide varieties of fields like bio-heat transfer, semiconductors, various cooling techniques, and natural phenomenon like oceanic currents and other important and relevant areas.

This chapter aims to cover all the relevant research papers about heat transfer published till 2018; few are there containing numerical and analytical aspects of heat transfer, while others are highlighted for its applications in engineering.

2. Free stream and flows over a surface

The chapters have been classified into categories like compressible and high-speed flows, externally influenced flows, flow related to films and interfaces, instable flow effects, flows with special fluid types, and flow related to reactions.

2.1 Effect of external surface

The effect of turbulence on free stream during heat transfer enhancement caused by the destruction of the viscous sublayer in the gaseous cavitation of CO_2-saturated water was recognized. The influence of roughness and wall temperature on the turbulent boundary layers was investigated [1, 2]. A model was developed to evaluate fluxes in urban boundary layers using the naphthalene sublimation technique [3].

2.2 Effect of geometry

Heat transfer enhancement is the process of improving the rate of heat deposition or removal on a surface. It is a subject of interest to the researchers as it results in savings in energy as well as cost. Heat transfer can be enhanced by using different types of swirl generators. Geometry plays a vital role in heat transfer enhancement. Transverse ribs with twisted tape and helical tape; axial rib with screw tape; and inclined limb in cylindrical dust have been studied for friction factor and Nusselt number [4–8]. Heat transfer augmentation techniques have been used to the study the effect of heat transfer and pressure drop due to insertion of twisted tape, inclined turbulator, corrugated tube with spring tape, diamond shape cylinder, wavy turbulator for short length and full length, center-trimmed twisted tape, flow around hexagonal cylinder, wavy channel, rhombus duct, square duct, and double pipe [9–20] as shown in **Figure 1**.

2.3 High-speed flow

A computational fluid dynamic (CFD) model has been developed to understand the hypersonic flow fields for reentry vehicles; facility was created for modeling

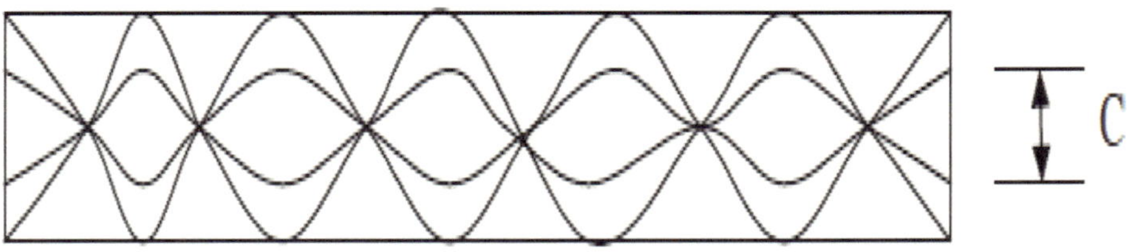

(a)

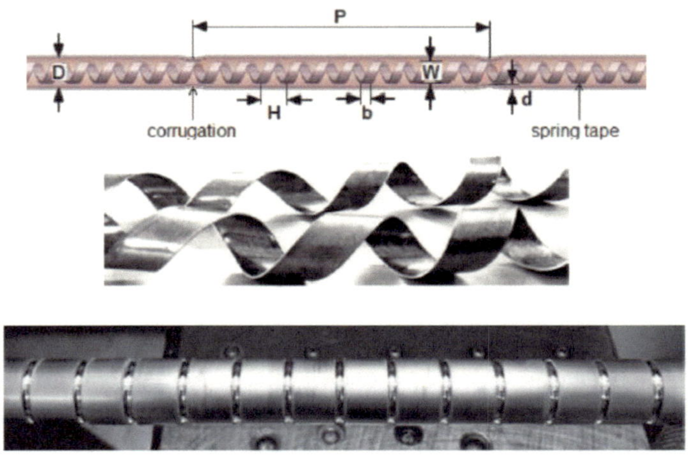

(b)

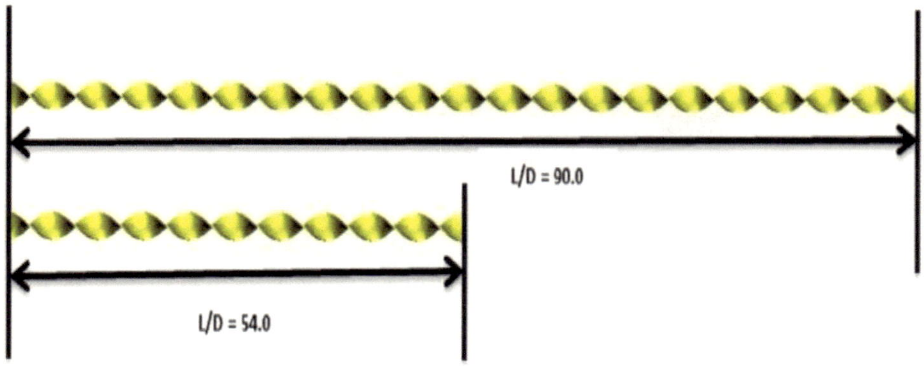
L/D = 90.0
L/D = 54.0

(c)

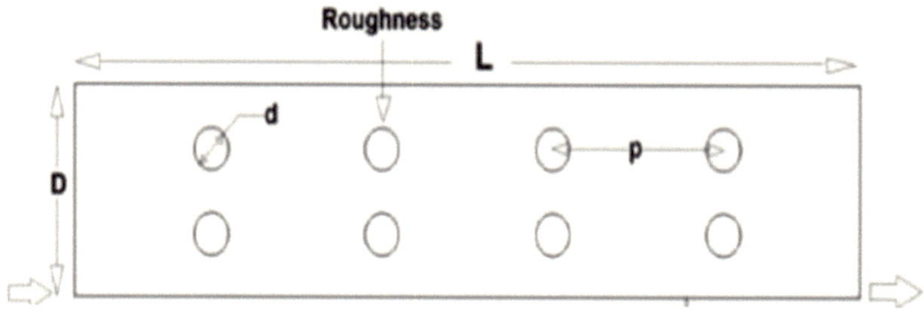
Roughness
L
d
p
D

(d)

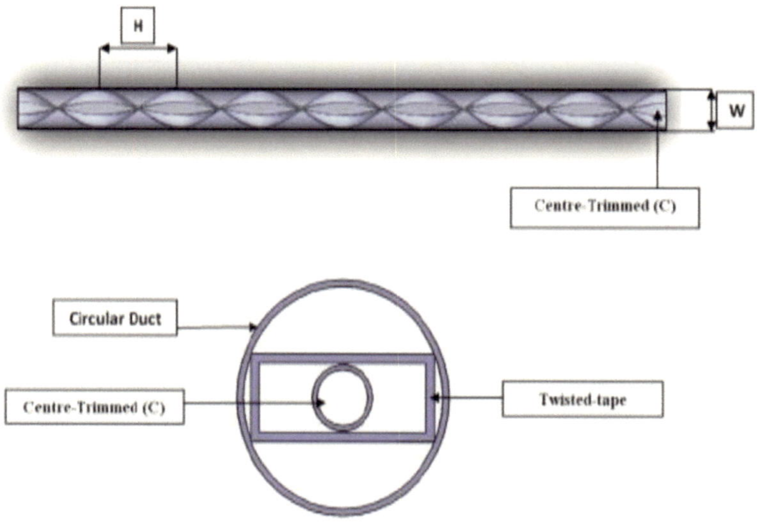
H
W
Centre-Trimmed (C)
Circular Duct
Centre-Trimmed (C)
Twisted-tape

(e)

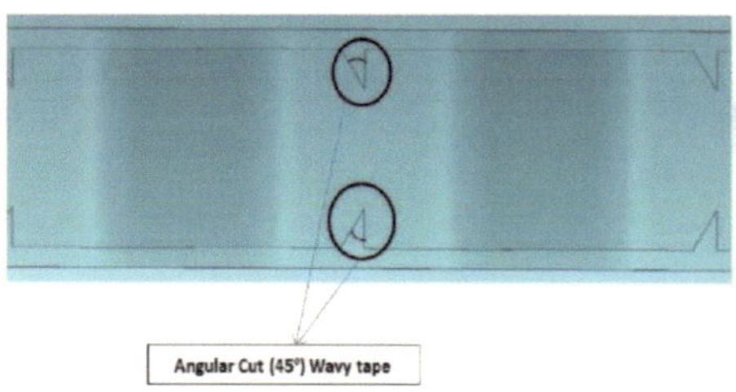
Angular Cut (45°) Wavy tape

(a)

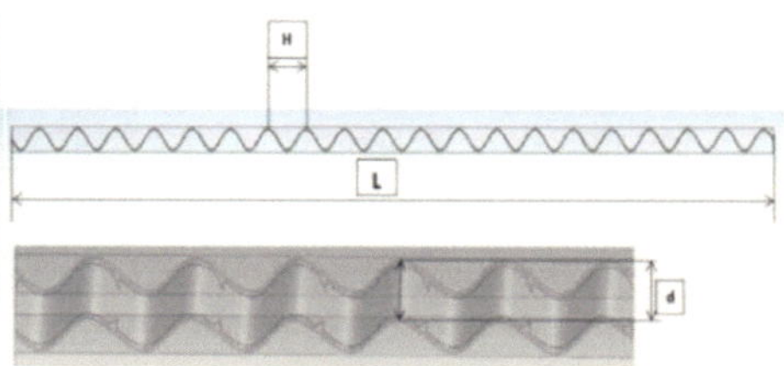
H
L
d

(b)

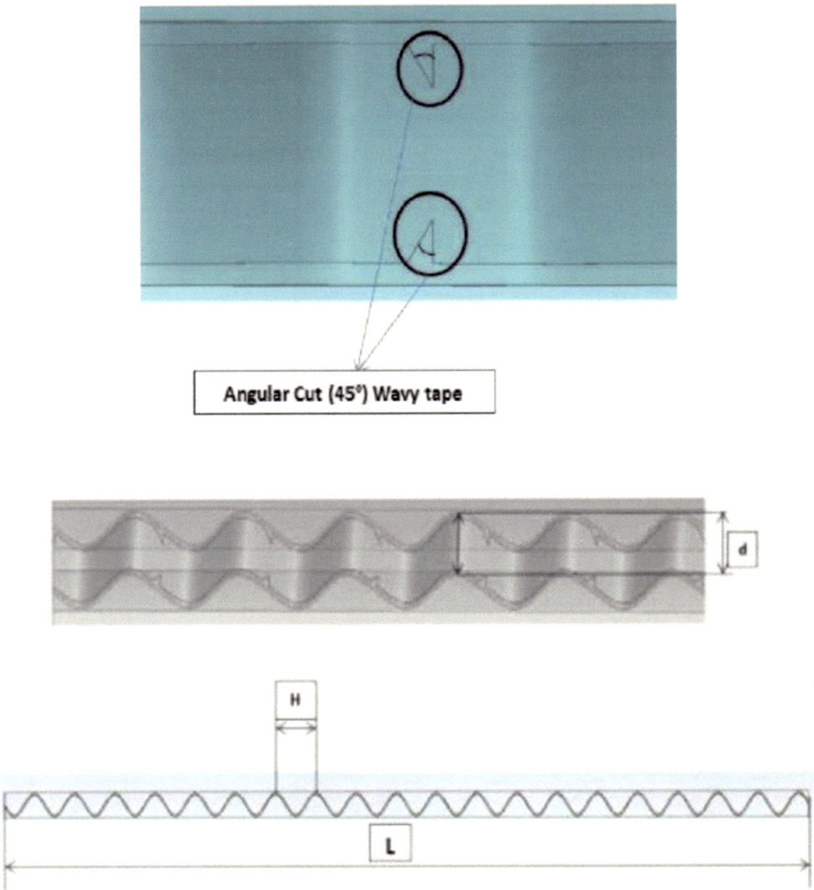

(f)

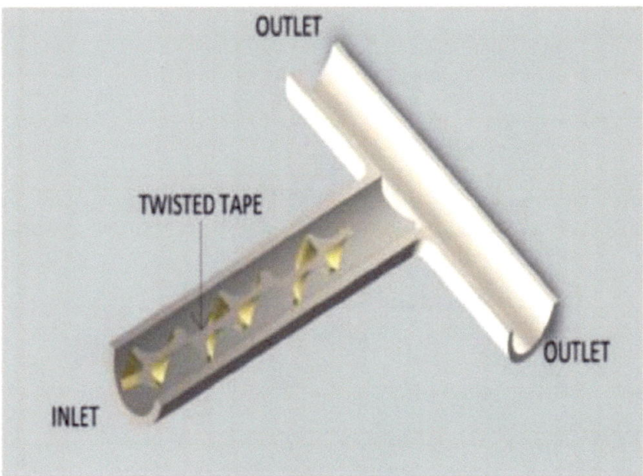

(g)

Figure 1.
Different types of vortex generators used for enhancement of heat transfer. (a) Center-cleared twisted tape [4]. (b) Spring tape insert [9]. (c) Twisted tape [10]. (d) Swirl generator [12]. (e) Twisted tape with clearance at the center [13]. (f) Wavy tape with angular cuts [14]. (g) Full-length twisted tape [16].

the projectile flight heating upon reentry. Simulation model for heat transfer due to convection and heat penetration was proposed [21], and comparative study has been conducted using of the European Atmospheric Reentry Demonstrator.

3. Channel flows

3.1 Straight wall passage

Initially, thermal characteristics in straight wall passages have been considered to analyze the heat transfer phenomenon in channel flows. Using the finite elements method, Nusselt number and friction factor were calculated for laminar regime. An investigation on laminar-turbulent transition inside a heated horizontal tube was conducted [22]. An analytical study for joule heating in a parallel plate channel with thermally developed flow has been conducted [23]. A circular tube was examined using various different conditions for viscous flow [24]. A novel method was developed for evaluating the Nusselt number for hydrodynamic flow conditions [25]. Horizontal, inclined channels and vertical plane passages were examined for mixed convective heat transfer [26, 27]. A prediction was presented for Nusselt number for the in-tube cooling of supercritical carbon dioxide [28].

3.2 Microscale heat transfer

The study of fine scale heat transfer was done with various channel configurations. 3D flow and heat transfer were examined in microchannels [29]. Theoretical analysis for heat transfer in laminar flow between two parallel plates separated by a very small space in micron range was conducted. The momentum and energy equations are solved for the hydraulic and fully developed thermal flow in the microchannel [30]. This method was also used to simulate rarefied gas flow and heat transfer in microchannels in a particular Knudsen number range [31]. Water was used as the working fluid in microchannel of rectangular shaped heat sinks, and computational studies were carried out [32]; also, their thermal performance was optimized minus water [33]. Convective heat transfer of fully developed flow both thermally and hydrodynamically in a rectangular microchannel is investigated [34]. A simulation model of low-power microchannel thermal reactor was presented [35]. Fractal branching used for the cooling of electronic chips was investigated [36]. Slotted microchannels were studied analytically on the basis of conduction and convection [37]. The performance of thermal fluid in a small capillary was studied experimentally [38].

3.3 Irregular geometries

A variety of papers covering numerous geometries have been taken into consideration in this section. Narrow-spaced fuel element configuration in multichannel was modeled numerically [39]. Rhombus and ellipse shape ducts were studied using Galerkin integral method [40, 41]. The heat transfer in a pin fin at the end of the wall was investigated [42]. The heat transfer in a milliscale thrust nozzle was studied numerically [43]. Viscous flow convections and heat transfer were studied in corrugated ducts [44]. For square ducts a combined study was undertaken to understand the thermal characteristics in different shapes [45]. Experimental study was done with regard to two-pass internal coolant passages in gas turbines [46]. An increase in heat transfer due to rolling and pitching action in swirling ducts was found experimentally [47]. Flow and heat transfer for metal honeycomb geometry was inspected [48]. The effect of viscous forced convection in branching ducts was studied [49].

4. Flow separation

A study showing the separation of energy in free shear zone was carried out, and the role of pressure in the flow separation was studied [50]. In vertical ducts and other cross sections, mixed convection was investigated numerically [51]. Three-dimensional flow studies for understanding the effect of step height were also available. Large eddy simulation (LES) was undertaken for the turbulent flow over a backward-facing step [52]. A laminar airfoil was taken, and adiabatic and heating conditions were investigated at modest subsonic Mach numbers [53]. Correlations among the heat transfer coefficients of dull-edged flat plates and square channels were studied [54]. A finite volume method was used to investigate the 2D natural convection in a heated cylinder, and the significance of aspect ratio, Prandtl number, and boundary conditions on thermal characteristics were studied [55].

5. Experimental methods

Precise measurement results in good outcome in every research work. In heat transfer, measurement plays a pivotal role in the analysis of thermal system. Even after the numerical modeling of a heat flow system, it is not possible to define all the parameters with full accuracy leading to the failure of many thermal systems. This includes most of the engineering devices such as spacecraft, cryogenic engines, satellites, etc. Modeling of a turbulent flow and transition zone is very complicated, and hence it is difficult to predict very accurately. For the accuracy and relevancy of data, precise measurement is required which gives rise to the development of precise system with better accuracy.

5.1 Heat transfer

Heat flux measurement is an important aspect for understanding the physics related to transport of heat; distinguish among conduction, convection, and radiation mechanism; analyze energy balance; derive material properties; and understand the flow regimes, etc. The physical and mathematical models are presented to investigate the evolution of surface waves for free-falling turbulent wavy films with varying Reynolds number [56]. Thermochromic liquid crystal (TLC) is one of the best techniques to visualize the transient heat flow over the surface. The change in color of TLC from red to green to blue helps understand the flow of heat flux over a surface in supersonic wind tunnel [57]. Luminescent coating is another option for measurement of heat flow over a surface. The method has been used to determine the heat flow in shorter duration of less than 10 ms hypersonic flow [58]. Heater foils can be used to measure the bulk temperature development in time-based heat flow measurement by using a simplified model in which temperature development has been characterized [59].

5.2 Temperature measurement

For the temperature measurement during rapid contact solidification at the surface of substrate, an interfacial temperature sensor of 1 μm diameter has been fabricated [60]. A telecentric objective has been used for the first time to phase out the dependence on the angles in color determination in fluid-based TLCs for precision measurement [61]. An acoustic thermal scan has been evaluated using numerical methodology to evaluate the spatial resolution [62].

5.3 Velocity measurement

A multiple hot-film sensor (MHFS) arrays were used to evaluate the skin friction along the surfaces of two-dimensional streamlined objects (circular cylinder) [63]. A high time resolution ultrasonic velocity profiler (UVP) system has been developed to determine 1D velocity profile on an ultrasonic beamline [64]. A numerical investigation [65] was conducted to determine the thermal response of hot wire for the measurement of sudden shift in the velocity in turbulent flow.

5.4 Miscellaneous

Bubble cluster pattern has been reported in the turbulent bubbly flow using rake of resistive flow and signal processing associated with it [66]. Air and liquid flows have been measured individually in two-phase air liquid flow [67]. Ultrasound Doppler velocimetry has been used to measure the thickness and velocity of the liquid film [68]. A novel pressure-sensitive paint (PSP) has been formulated and used for pressure measurement in cryogenic wind tunnel [69]. A high-sensitivity thermal conductivity detector has been developed from different materials which can be used in the diagnosis of fault in transformer, oil exploration, etc. [70].

6. Phase change

This part of the chapter deals with melting and freezing of materials. The section is divided into several subsections such as phase change materials (PCMs); formation of ice and its melting; melting and freezing of radial objects; melting and solidification of metals, nonmetals, and composites; crystallization; and globule, spray, and plunge cooling for better understanding.

6.1 Phase change materials

The inability to recover latent heat after super cooling of PCMs has been pointed out, and the method to recover latent heat has been discussed [71]. The melting mechanism of PCMs in magnetic field in low-gravity atmosphere has been discussed [72]. Other works include fabrication of carbon brushes which can be used to enhance the thermal conductivity in phase change materials, the role of ultrasonic vibration on melting characterization of PCMs, and detailed examination of solid liquid phase change heat flow enhancement.

6.2 Formation of ice and its melting

Researches on the formation of ice and its melting include the thermal behavior of ice under constant heat flow per unit area and melt removal, melting of ice using natural convection, ice making by cooling water-oil emulsion with stirring, and numerical simulation of melting of ice in water under the influence of natural convection and cooling effect produced by melting of ice [73–75].

6.3 Melting and freezing of radial objects

A lot of work has been presented on phase change in radial objects such as sphere, cylinders, and slabs. A mathematical model using numerical analysis has been developed to study the melting process of PCMs in sphere [76]. A novel packed bed of spheres has been developed using graphite/PCM composite for increasing the thermal conductivity which resulted in reduction in melting and freezing time significantly [77].

6.4 Melting and solidification of metals, non-metals, and composites

It has been observed that the supercooling properties of sodium acetate trihydrate can be improved by addition of nano-Cu [78]. A simulation model of melting and solidification of PCM in metallic porous foam has been investigated in the heat exchanger. The Hunt-Trivedi model has been used to simulate the solidification process of AISI 304 stainless steel [79].

6.5 Crystal growth

The crystal growth involves controlled growth of microstructure using optical heating, modeling of mass crystallization in magma chamber, effect of crystal growth on solute distribution, and simulation of crystal growth for binary melting process [80, 81].

6.6 Globules, spray, and plunge cooling

This subsection deals with the deposition of metal droplets on the premolten pool and wavy surfaces, numerical analysis using FEM to study fluid mechanics and heat transfer of solder droplet on flat surface, numerical analysis of microdroplet deposition over a novel micromanufacturing process, and utilization of impulse atomization technique to produce controlled size droplets [82–84].

7. Numerical methods

Numerical methods are used to develop the mathematical models to solve complex numerical problems. The technique is used widely in research for modeling and optimization of the physical work which otherwise required rigorous work. The research work done in the field of heat transfer using numerical methods has been depicted in this section.

7.1 Heat conduction

A hybrid 3D model has been developed for the analysis of transient heat conduction in a functionally graded material (FGM) using generalized finite difference method [85], Cattaneo-Vernotte model (CV model) was used to develop numerical simulation of non-Fourier heat conduction for a fin attached to a microelectronic surface [86], Galerkin-vector theory and numerical method are used to develop a mathematical model to study heat conduction in nonhomogenous materials [87], and heat conduction model was developed using numerical methods to understand the flow of heat in the granular materials [88].

7.2 Inverse analysis

Systematic and local error has been identified using WKB method through numerical analysis [89], numerical inverse Laplace transform was used to solve nonlinear differential Equation [90], and numerical inverse method has been developed to extract heat flux in heat-sensitive coating region [91].

7.3 Fluid flow

The lattice Boltzmann method is used for simulation model of non-Newtonian fluid flow, two fluid method, and discrete particle method used for simulating the gas-solid flow of rough particles. A CFD model can be used effectively to study the hydrofluidization freezing, and a numerical simulation of fluid flow with thermal hydraulic mechanical coupling method on an uneven surface was developed [89].

7.4 Turbulent flow

Numerical methods can also be utilized to predict the turbulent flow. k-ϵ and LES model were used to study turbulent flow field around rows of tree and building, turbulence in flow field and temperature can be predicted, renormalization is used to determine the eddy diffusion in turbulence flow, intermittency model was developed for studying the laminar boundary transition at supersonic and hypersonic condition, and LES is used to forecast the heat transfer coefficient and blade metal temperature [92].

8. Heat exchanger and thermosyphons

8.1 Applications

The sheer variety of heat transfer operations has been demonstrated by a number of researchers in their works dealing with thermoacoustic and thermoelectric devices, rotating heat exchangers, commercial blood oxygenators, soil and deep bore heat exchangers, space craft radiators, and pressurized bubble columns.

8.2 Enhancement of heat transfer

The procedure to ease heat transfer has been stated by many researchers. The fin technology of extension is quite prevalent in the recent times. An investigation was carried out with fin tubes using liquid crystal display technology and plate finned tube exchanger by infrared thermal imaging, and performance measurement has been reported for a finned tube surface and annular fins. Fins having curly surfaces are examined for humid airflow. In addition to this film-wise condensation on plane low finned tubes, transient conduction in a fin, performance of extruded-serrated and extruded-finned tube bundles, and the features of a multi-pass heat exchanger have also been reported.

8.3 Microscale heat transfer

A number of applications now employ miniaturization of heat transfer devices: micro-heat pipe arrays, electronic cooling, microturbine, evaporation and boiling in microfin, microheat pipes, microscale temperature measurements, and modeling of microchannel flows [93].

8.4 Effect of fouling

An investigation has been done to study the effect of gas-side fouling in cross flow. Calcium carbonate fouling effect was studied with a microscale image; mineral fouling in extended tube heat exchangers was studied; the use of polyacrylic acid as anti-scaling and antifouling agent was studied [94–96].

8.5 Systems based on thermosyphon

Thermosyphons found applications in a variety of heat transfer complications such as space radiators and cooling of structures, solar water heaters, nuclear reactors and system based on geothermal energy, evaporators, preheaters, tiny heat pipes used to cool PC, laptops, and other electronic components [97–100].

9. Heat transfer: general applications

The relationship between the parameters of a fluidized bed and the heat transfer to a body engulf in it [101]. The SIMPLE algorithm [102] was used to simulate a blast furnace, and a relationship has been developed for the heat transfer coefficients on extension walls and hydrowalls of the boilers [103]. A porous radiant recirculated burner (PRRB) concept is developed to reduce losses due to open-flame combustion [104]. Leong studied the effect of latent heat of fusion on thin plates and numerical analysis of temperature change in biscuits using Monte Carlo (MC) method [105].

Multiple papers investigated the thermohydraulics of the cooling flow in nuclear reactors. A model was developed to study heat flux for low flow rates [106]. Interwrapper flow was studied, and its effects were analyzed numerically [107]. In the case of ceramic-coated turbine blades, the heat transfer coefficient does not significantly affect metal temperatures when thermal radiation is in the picture.

10. Insolation

10.1 Solar radiation

Various perspectives to evaluate solar data using modified modeling have been conducted by researchers. A new correlation between sunshine duration and radiation on the surface of the earth has been derived by Suehrcke [108]. It was found that the correlation is very well established for average value. A correction factor was proposed by Muneer [109] for calibrating the shadow band pyranometer. A model was using upper air humidity to estimate global solar radiation [110].

10.2 Solar air heater

Numerical solutions were also developed for absorbers in a porous medium, Nusselt number and Reynolds-Rayleigh number correlations for natural convection in an open-ended rectangular channel and models for solar air heater with fins [111]. Collector efficiency was predicted in a simplified manner.

10.3 Solar water heaters

Novelties in the design of solar water heating application are presented in this subsection. Fourier transform technique has been used to estimate the heat transfer and efficiency of a flat solar plate collector [112]. Double-sided flat plate collector was used to experimentally investigate the reduction in heat losses in comparison to conventional solar collector [113]. An experimental investigation on ICS solar water heart with compound parabolic concentrating integral collector storage system was designed and tested [114]. A 2D concentrator was developed aiming to store solar energy [115].

11. Plasma heat transfer and MHD

11.1 Investigation and application

In this section, heat transfer in thermal plasma reactor for nanoparticle synthesis has been investigated through different models [116]. A 3D model of heat transfer in thermal plasma system has been developed to show 3D effect of carrier gas [117]; effects of nucleation temperature were investigated by radio frequency; 2D numerical simulation was developed to show flow and heat transfer in argon gas plasma, temperature gradient, velocity, and concentration to study the nitridation of $MoSi_2$ which was carried in thermal plasma reactor; and numerical simulation model was developed to show the effect of radial injection of gas (with and without swirl) on flow and temperature field [118]. Plasma induced between two electrodes with and without swirl has been investigated for heat transfer with fluid flow [119].

11.2 Magnetohydrodynamics (MHD)

A simple monoenergetic operator and the Bhatnagar-Gross-Krook model were presented to estimate heat transfer in a rare gas between parallel plates [120]. A mathematical model of 2D magnetohydrodynamic Prandtl fluid flow over a sheet is examined [121], 3D magnetohydrodynamic Cauchy problem has been investigated [122], 2D pseudo-steady compressible magnetohydrodynamic system is studied for expansion of gas in vacuum [123], bio-convection flow of nanofluid is studied in magnetic field [124], and numerical analysis of MHD flow over vertical rotating cone is investigated. Also, the volume of fluid method is used to investigate the MHD of incompressible flow, and MHD fluid behavior is studied for flow and heat transfer [125].

12. Conclusion

In this review article, an effort has been made to study the recent development in the field of heat transfer enhancement. A lot of experimental and numerical research have been done to study the aspect of heat transfer in different fields such as channel flow, crystal growth, heat exchangers, thermosyphons, phase change materials, temperature and velocity measurement, solar energy, etc. The effect of geometry such as channel modification through inserts, roughness, etc. and external power such as magnetic field, electric field, ultrasound, etc. on the thermal performance and augmentation of heat transfer has been studied. In addition to this, the lattice Boltzmann method, WKB method, numerical inverse method, k-epsilon, Cattaneo-Vernotte model, Hunt-Trivedi model, and LES model have been studied for different heat transfer applications. Overall this review gives a full-scale summary of heat transfer applications.

Author details

Suvanjan Bhattacharyya[1*], Devendra K. Vishwakarma[2], Sanghati Roy[2], Ranjib Biswas[3] and Mohammad Moghimi Ardekani[4]

1 Center for Renewable Energy and Environment Development (CREED), Department of Mechanical Engineering, Birla Institute of Technology and Science, Pilani, Vidyavihar, Rajasthan, India

2 Department of Mechanical Engineering, Birla Institute of Technology and Science, Pilani, Vidyavihar, Rajasthan, India

3 Department of Mechanical Engineering, MCKV Institute of Engineering, Howrah, West Bengal, India

4 Department of Design and Engineering, Staffordshire University, Stoke-on-Trent, UK

*Address all correspondence to: suvanjan.bhattacharyya@pilani.bits-pilani.ac.in

References

[1] Lee SY, Choi YD. Turbulence enhancement by ultrasonically induced gaseous cavitation in the CO_2 saturated water. KSME Journal. 2002;**16**(2):246-254

[2] Avelino MR, Freire APS. On the displacement in origin for turbulent boundary layers subjected to sudden changes in wall temperature and roughness. International Journal of Heat and Mass Transfer. 2002;**45**(15):3143-3153

[3] Barlow JF, Belcher SE. A wind tunnel model for quantifying fluxes in the urban boundary layer. Boundary-Layer Meteorology. 2002;**104**(1):131-150

[4] Bhattacharyya S, Saha S, Saha SK. Laminar flow heat transfer enhancement in a circular tube having integral transverse rib roughness and fitted with centre-cleared twisted-tap. Experimental Thermal and Fluid Science. 2012;**44**:727-735

[5] Bhattacharyya S, Saha SK. Thermohydraulics of laminar flow through a circular tube having integral helical rib roughness and fitted with centre-cleared twisted-tape. Experimental Thermal and Fluid Science. 2012;**42**:154-162

[6] Saha SK, Bhattacharyya S, Pal PK. Thermohydraulics of laminar flow of viscous oil through a circular tube having integral axial rib roughness and fitted with center-cleared twisted-tape. Experimental Thermal and Fluid Science. 2012;**41**:121-129

[7] Bhattacharyya S, Chattopadhyay H, Benim AC. 3D CFD simulation on heat transfer enhancement in turbulent channel flow with twisted tape inserts. Progress in Computational Fluid Dynamics, An International Journal. 2017;**17**(3):193-197

[8] Bhattacharyya S, Chattopadhyay H, Benim AC. Numerical investigation on heat transfer in a circular tube with inclined ribs. Progress in Computational Fluid Dynamics, An International Journal. 2017;**17**(6):390-396

[9] Bhattacharyya S, Benim AC, Chattopadhyay H, Banerjee A. Experimental investigation of the heat transfer performance of a corrugated tube with spring tape turbulator inserts. Experimental Heat Transfer. 2019;**32**(5):411-425

[10] Bhattacharyya S. Experimental study on effect of heat transfer enhancement of heat exchanger tube inserted with short length spring tapes. Iranian Journal of Science and Technology Transactions of Mechanical Engineering. DOI: 10.1007/s40997-018-0251-0

[11] Bhattacharyya S, Benim AC, Chattopadhyay H, Banerjee A. Experimental and numerical analysis of forced convection in a twisted tube. Journal of Thermal Science. 2019;**23**(4):1043-1052

[12] Bhattacharyya S, Chattopadhyay H. Experimental investigation on heat transfer enhancement by swirl generators in a solar air heater duct. International Journal of Heat and Technology. 2016;**34**(2):191-196

[13] Bhattacharyya S, Chattopadhyay H, Bandhopadhyay S. Numerical study on heat transfer enhancement through a circular duct fitted with centre-trimmed twisted tape. International Journal of Heat and Technology. 2016;**34**(3):401-406

[14] Bhattacharyya S, Chattopadhyay H, Benim AC. Heat transfer enhancement of laminar flow of ethylene glycol through a square channel fitted with

angular cut wavy strip. Procedia Engineering. 2016;**157**:19-28

[15] Bhattacharyya S, Chattopadhyay H, Swami A, Uddin MK. Convective heat transfer enhancement and entropy generation of laminar flow of water through a wavy channel. International Journal of Heat and Technology. 2016;**34**(4):727-733

[16] Bhattacharyya S, Chattopadhyay H, Haldar A. Design of twisted tape turbulator at different entrance angle for heat transfer enhancement in a solar heater. Beni-Suef University Journal of Basic and Applied Sciences. 2018;**7**(1):118-126

[17] Nandi TK, Bhattacharyya S, Gobinda Das S, Chattopadhyay H. Thermohydraulic transport characteristics in wavy microchannel under pulsating inlet flow condition. Chemical Engineering Transactions. 2017;**62**:271-276

[18] Bhattacharyya S, Das S, Sarkar A. Numerical simulation of flow and heat transfer around hexagonal cylinder. International Journal of Heat and Technology. 2017;**35**(2):360-363

[19] Bhattacharyya S, Khan AI, Maity DK, Pradhan S. Hydrodynamics and heat transfer of turbulent flow around a rhombus cylinder. Chemical Engineering Transactions. 2017;**62**:373-378

[20] Bhattacharyya S, Sarkar A, Das S, Mullick A. Computational studies of heat transfer enhancement in a circular wavy micro channel. Chemical Engineering Transactions. 2017;**62**:361-366

[21] Meese EA, Norstrud H. Simulation of convective heat flux and heat penetration for a spacecraft at re-entry. Aerospace Science and Technology. 2002;**6**(3):185-194

[22] Koizumi H. Laminar-turbulent transition behavior of fully developed air flow in a heated horizontal tube. International Journal of Heat and Mass Transfer. 2002;**45**(5):937-949

[23] Lahjomri J, Oubarra A, Alemany A. Heat transfer by laminar Hartmann flow in thermal entrance region with a step change in wall temperatures: The Graetz problem extended. International Journal of Heat and Mass Transfer. 2002;**45**(5):1127-1148

[24] Mahulikar SP, Tso CP. A new classification for thermal development of fluid flow in a circular tube under laminar forced convection. Proceedings of the Royal Society of London Series A: Mathematical, Physical and Engineering Sciences. 2002;**458**(2019):669-682

[25] Nickolay M, Martin H. Improved approximation for the Nusselt number for hydrodynamically developed laminar flow between parallel plates. International Journal of Heat and Mass Transfer. 2002;**45**(15):3263-3266

[26] Ozsunar A, Baskaya S, Sivrioglu M. Experimental investigation of mixed convection heat transfer in a horizontal and inclined rectangular channel. Heat and Mass Transfer. 2002;**38**(3):271-278

[27] Wang J, Li J, Jackson JD. Mixed convection heat transfer to air flowing upwards through a vertical plane passage: Part 3. Chemical Engineering Research and Design. 2002;**80**(A3):252-260

[28] Pitla SS, Groll EA, Ramadhyani S. New correlation to predict the heat transfer coefficient during in-tube cooling of turbulent supercritical CO_2. International Journal of Refrigeration - Revue internationale du Froid. 2002;**25**(7):887-895

[29] Toh KC, Chen XY, Chai JC. Numerical computation of fluid flow

and heat transfer in microchannels. International Journal of Heat and Mass Transfer. 2002;**45**(26):5133-5141

[30] Zhu X, Xin MD, Liao Q. Analysis of heat transfer between two unsymmetrically heated parallel plates with microspacing in the slip flow regime. Microscale Thermophysical Engineering. 2002;**6**(4):287-301

[31] Wang X, Wang QW, Tao WQ, Zheng P. Simulation of rarefied gas flow and heat transfer in microchannels. Science in China, Series E Technological Sciences. 2002;**45**(3):321-327

[32] Qu WL, Mudawar I. Analysis of three-dimensional heat transfer in micro-channel heat sinks. International Journal of Heat and Mass Transfer. 2002;**45**(19):3973-3985

[33] Ryu JH, Choi DH, Kim SJ. Numerical optimization of the thermal performance of a microchannel heat sink. International Journal of Heat and Mass Transfer. 2002;**45**(13):2823-2827

[34] Tunc G, Bayazitoglu Y. Heat transfer in rectangular microchannels. International Journal of Heat and Mass Transfer. 2002;**45**(4):765-773

[35] Erickson D, Li DQ. Numerical simulations of a low power microchannel thermal cycling reactor. International Journal of Heat and Mass Transfer. 2002;**45**(18):3759-3770

[36] Chen YP, Cheng P. Heat transfer and pressure drop in fractal tree-like microchannel nets. International Journal of Heat and Mass Transfer. 2002;**45**(13):2643-2648

[37] Leont'ev AI, Polyakov AF. Formulation and solution of the problem on convection-conduction heat transfer in a system of slotted microchannels at a uniform temperature of the frame. High Temperature. 2002;**40**(4):577-585

[38] Celata GP, Cumo M, Guglielmi M, Zummo G. Experimental investigation of hydraulic and single-phase heat transfer in 0.130-mm capillary tube. Microscale Thermophysical Engineering. 2002;**6**(2):85-97

[39] Huang J, Wang QW, Tao WQ. Numerical simulation of turbulent flow and heat transfer in multi-channel, narrow-gap fuel element. Engineering Computations (Swansea, Wales). 2002;**19**(3-4):327-345

[40] Lee YM, Lee PC. Heat transfer coefficients of laminar flow in a rhombic duct with constant wall temperature. Numerical Heat Transfer: Part A Applications. 2002;**42**(3):285-296

[41] Sakalis VD, Hatzikonstantinou PM, Kafousias N. Thermally developing flow in elliptic ducts with axially variable wall temperature distribution. International Journal of Heat and Mass Transfer. 2002;**45**(1):25-35

[42] Hwang JJ, Lui CC. Measurement of endwall heat transfer and pressure drop in a pin-fin wedge duct. International Journal of Heat and Mass Transfer. 2002;**45**(4):877-889

[43] Alexeenko AA, Levin DA, Gimelshein SF, Collins RJ, Markelov GN. Numerical simulation of high-temperature gas flows in a millimeter-scale thruster. Journal of Thermophysics and Heat Transfer. 2002;**16**(1):10-16

[44] Niu JL, Zhang LZ. Heat transfer and friction coefficients in corrugated ducts confined by sinusoidal and arc curves. International Journal of Heat and Mass Transfer. 2002;**45**(3):571-578

[45] Wang LB, Wang QW, He YL, Tao WQ. Experimental and numerical study of developing turbulent flow and heat transfer in convergent/divergent square ducts. Heat and Mass Transfer. 2002;**38**(4-5):399-408

[46] Chanteloup D, Juaneda Y, Bolcs A. Combined 3-D flow and heat transfer measurements in a 2-pass internal coolant passage of gas turbine airfoils. Journal of Turbomachinery-Transactions of the ASME. 2002;**124**(4):710-718

[47] Chang SW, Zheng Y. Enhanced heat transfer with swirl duct under rolling and pitching environment. Journal of Ship Research. 2002;**46**(3):149-166

[48] Brautsch A, Griffin T, Schlegel A. Heat transfer characterization of support structures for catalytic combustion. International Journal of Heat and Mass Transfer. 2002;**45**(15):3223-3231

[49] El-Shaboury AMF, Soliman HM, Ormiston SJ. Laminar forced convection in two-dimensional equalsided and reduced branching ducts. Numerical Heat Transfer: Part A Applications. 2002;**42**(5):487-512

[50] Han B, Goldstein RJ, Choi HG. Energy separation in shear layers. International Journal of Heat and Mass Transfer. 2002;**45**(1):47-55

[51] Cheng CH, Huang SY, Aung W. Numerical predictions of mixed convection and flow separation in a vertical duct with arbitrary cross section. Numerical Heat Transfer: Part A Applications. 2002;**41**(5):491-514

[52] Nie JH, Armaly BF. Three-dimensional convective flow adjacent to backward-facing step-effects of step height. International Journal of Heat and Mass Transfer. 2002;**45**(12):2431-2438

[53] Stock HW. Wind tunnel-flight correlation for laminar wings in adiabatic and heating flow conditions. Aerospace Science and Technology. 2002;**6**(4):245-257

[54] Sun ZF. Correlations for mass transfer coefficients over blunt boards based on modified boundary layer theories. Chemical Engineering Science. 2002;**57**(11):2029-2033

[55] Roychowdhury DG, Das SK, Sundararajan T. Numerical simulation of natural convective heat transfer and fluid flow around a heated cylinder inside an enclosure. Heat and Mass Transfer. 2002;**38**(7-8):565-576

[56] Ye XM, Yan WP, Jiang ZY, Li CX. Hydrodynamics of free-falling turbulent wavy films and implications for enhanced heat transfer. Heat Transfer Engineering. 2002;**23**(1):48-60

[57] Mee DJ, Chiu HS, Ireland PT. Techniques for detailed heat transfer measurements in cold supersonic blowdown tunnels using thermochromic liquid crystals. International Journal of Heat and Mass Transfer. 2002;**45**(16):3287-3297

[58] Hubner JP, Carroll BF, Schanze KS. Heat-transfer measurements in hypersonic flow using luminescent coating techniques. Journal of Thermophysics and Heat Transfer. 2002;**16**(4):516-522

[59] von Wolfersdorf J. Bulk temperature development in transient heat transfer measurements using heater foils. Journal of Heat Transfer-Transactions of the ASME. 2002;**124**(5):982-985

[60] Wang W, Qiu HH. Interfacial thermal conductance in rapid contact solidification process. International Journal of Heat and Mass Transfer. 2002;**45**(10):2043-2053

[61] Gunther A, von Rohr PR. Influence of the optical configuration on temperature measurements with fluid dispersed TLCs. Experiments in Fluids. 2002;**32**(5):533-541

[62] Bograchev KM, Passechnik VI. Evaluation of the spatial resolution of

passive acoustic thermal tomography. Acoustical Physics. 2002;**48**(4):406-411

[63] Desgeorges O, Lee T, Kafyeke F. Multiple hot-film sensor array calibration and skin friction measurement. Experiments in Fluids. 2002;**32**(1):37-43

[64] Ozaki Y, Kawaguchi T, Takeda Y, Hishida K, Maeda M. High time resolution ultrasonic velocity profiler. Experimental Thermal and Fluid Science. 2002;**26**(2-4):253-258

[65] Durst F, Shi JM, Breuer M. Numerical prediction of hot-wire corrections near walls. Journal of Fluids Engineering, Transactions of the ASME. 2002;**124**(1):241-250

[66] Salesse A, de Tournemine AL, Roig V. Development of bubble cluster detection and identification method. Experimental Thermal and Fluid Science. 2002;**26**(2-4):163-171

[67] Dong W, Hu LZ. Gas-liquid two-phase flow measurement using ESM. Experimental Thermal and Fluid Science. 2002;**26**(6-7):827-832

[68] Zhaojun F, Jing C, Shitao S. Investigating the liquid film characteristics of gas-liquid swirling flow using ultrasound Doppler velocimetry. AICHE Journal. 2017;**63**:2348-2357

[69] Asai K, Amao Y, Iijima Y, Okura I, Nishide H. Novel pressure-sensitive paint for cryogenic and unsteady wind-tunnel testing. Journal of Thermophysics and Heat Transfer. 2002;**16**(1):109-115

[70] Wu YE, Chen K, Chen CW, Hsu KH. Fabrication and characterization of thermal conductivity detectors (TCDs) of different flow channel and heater designs. Sensors and Actuators A: Physical. 2002;**100**(1):37-45

[71] Zhao Y, Zhang X, Xu X, Zhang S. Research progress in nucleation and supercooling induced by phase change materials. Journal of Energy Storage. 2020;**27**:101156

[72] Asako Y, Goncalves E, Faghri M, Charmchi M. Numerical solution of melting processes for fixed and unfixed phase change material in the presence of magnetic field-simulation of low-gravity environment. Numerical Heat Transfer: Part A Applications. 2002;**42**(6):565-583

[73] Tsuchida D, Kang C, Okada M, Matsumoto K, Kawagoe T. Ice formation process by cooling water-oil emulsion with stirring in a vessel. International Journal of Refrigeration - Revue internationale du Froid. 2002;**25**(2):250-258

[74] Bansal H, Niktriyuk P. Arbitrary shaped ice particle melting under the influence of natural convection. AICHE Journal. 2017;**63**:3158-3176

[75] Yeh HM, Ho CD, Wang WP. Cool thermal discharge by melting ice and producing chilled air. Heat Transfer Engineering. 2002;**23**(1):61-68

[76] Wei L, Sui-Guang L, Shikai G, Yunhao W, Xueling L. Numerical study on melt fraction during melting of phase change material inside a sphere. International Journal of Hydrogen Energy. 2017;**42**(29):18232-18239

[77] Al-Shannaq R, Young B, Farid M. Cold energy storage in a packed bed of novel graphite/PCM composite spheres. Energy. 2019;**171**(15):296-305

[78] Wenlong C, Yanping Y, Liangliang S, Xiaoling C, Xiaojiao Y. Experimental studies on the supercooling and melting/freezing characteristics of nano-copper/sodium acetate trihydrate composite phase change materials. Renewable Energy. 2016;**99**:1029-1037

[79] Esapour M, Hamzehnezhad A, Ali Rabienaaj A, Darzi MJ. Melting and solidification of PCM embedded in porous metal foam in horizontal multi-tube heat storage system. Energy Conversion and Management. 2018;**171**:398-410

[80] Bazarov LS, Gordeeva VI, Shevchenko VS, Petrushin EI. Experimental modeling of mass crystallization processes in the volume of a flat magma chamber. Petrology. 2002;**10**(5):469-478

[81] Fukui K, Maeda K. Effects of crystal growth rate and heat and mass transfer on solute distribution. Chemical Engineering Science. 2002;**57**(15):3133-3140

[82] Butty V, Poulikakos D, Giannakouros J. Three dimensional presolidification heat transfer and fluid dynamics in molten micro droplet deposition. International Journal of Heat and Fluid Flow. 2002;**23**(3):232-241

[83] Haferl S, Poulikakos D. Transport and solidification phenomena in molten microdroplet pileup. Journal of Applied Physics. 2002;**92**(3):1675-1689

[84] Henein H. Single fluid atomization through the application of impulses to a melt. Materials Science & Engineering, A: Structural Materials - Properties, Microstructure and Processing. 2002;**326**(1):92-100

[85] Wenzhen Q, Chia-Ming F, Yaoming Z. Analysis of three dimensional heat conduction in functionally graded materials by using a hybrid numerical method. International Journal of Heat and Mass Transfer. 2019;**145**:118771

[86] Yi L, Ling L, Yuwen Z. Numerical simulation of non-Fourier heat conduction in fins by lattice Boltzmann method. Applied Thermal Engineering. 2020;**145**:114670

[87] Xiujiang S, Qian W, Liqin W. New Galerkin-vector theory and efficient numerical method for analyzing steady-state heat conduction in inhomogeneous bodies subjected to a surface heat flux. Applied Thermal Engineering. 2019;**161**:113838

[88] Jun H, Quansheng L, Zhijun W, Xiangyu X. Modelling transient heat conduction of granular materials by numerical manifold method. Engineering Analysis with Boundary Elements. 2018;**86**:45-55

[89] Ying X, Zhinxue S, Zhuang L, Jun Y, Dongyan F, Kelvin B, et al. Numerical simulation of fluid flow and heat transfer in EGS with thermal-hydraulic-mechanical coupling method based on a rough fracture model. Energy Procedia. 2019;**158**:6038-6045

[90] Mathey P, Jullien P. Numerical analysis of a WKB inverse method in view of index profile reconstruction in diffused waveguides. Optics Communications. 1996;**122**(4-6):127-134

[91] Rani D, Mishra V. Numerical inverse Laplace transform based on Bernoulli polynomials operational matrix for solving nonlinear differential equations. Results in Physics. 2020;**16**:102836

[92] Merci B, Dick E, Vierendeels J, De Langhe C. Determination of epsilon at inlet boundaries. International Journal of Numerical Methods for Heat and Fluid Flow. 2002;**12**(1):65-80

[93] Medic G, Durbin PA. Toward improved prediction of heat transfer on turbine blades. Journal of Turbomachinery-Transactions of the ASME. 2002;**124**(2):187-192

[94] Wand L, Song F. Development of an intermittency equation for the modeling

of the supersonic/hypersonic boundary layer flow transition. Flow, Turbulence and Combustion. 2011;**87**(1):165-187

[95] Tafti DK, He L, Nagendra K. Large eddy simulation for predicting turbulent heat transfer in gas turbines. Philosophical Transactions. Series A, Mathematical, Physical, and Engineering Sciences. 2014;**372**(2022):20130322

[96] Gui-Qing W, He-Ping L, Cheng-Yu B, Xi C. Modeling of the heat transfer and flow features of the thermal plasma reactor with counter-flow gas injection. International Journal of Heat and Mass Transfer. 2009;**52**(3-4):760-766

[97] Wang CC, Lo J, Lin YT, Wei CS. Flow visualization of annular and delta winlet vortex generators in fin-and-tube heat exchanger application. International Journal of Heat and Mass Transfer. 2002;**45**(18):3803-3815

[98] Wang YX, Peterson GP. Optimization of micro heat pipe radiators in a radiation environment. Journal of Thermophysics and Heat Transfer. 2002;**16**(4):537-546

[99] Schaller M, Hoffmann KH, Rivero R, Andresen B, Salamon P. The influence of heat transfer irreversibilities on the optimal performance of diabatic distillation columns. Journal of Non-Equilibrium Thermodynamics. 2002;**27**(3):257-269

[100] Bojic M, Lukic N. Controlling evaporative three finger thermosyphon. Energy Conversion and Management. 2002;**43**(5):709-720

[101] Al-Busoul MA. Bed-to-surface heat transfer in a circulating fluidized bed. Heat and Mass Transfer. 2002;**38**(4-5):295-299

[102] de Castro JA, Nogami H, Yagi J. Three-dimensional multiphase mathematical modeling of the blast furnace based on the multifluid model. ISIJ International. 2002;**42**(1):44-52

[103] Dutta A, Basu P. Overall heat transfer to water walls and wing walls of commercial circulating fluidized-bed boilers. Journal of the Institute of Energy. 2002;**75**(504):85-90

[104] Jugjai S, Rungsimuntuchart N. High efficiency heat-recirculating domestic gas burners. Experimental Thermal and Fluid Science. 2002;**26**(5):581-592

[105] Leong KC, Lu GQ, Rudolph V. The effect of latent heat of fusion on heat transfer in fluidized-bed coating of thin plates. Chemical Engineering and Processing. 2002;**41**(7):567-576

[106] Caro-Corrales J, Cronin K, Abodayeh K, Gutierrez-Lopez G, Ordorica-Falomir C. Analysis of random variability in biscuit cooling. Journal of Food Engineering. 2002;**54**(2):147-156

[107] Holowach MJ, Hochreiter LE, Cheung FB, Aumiller DL. Critical heat flux during reflood transients in small-hydraulic-diameter geometries. Nuclear Technology. 2002;**140**(1):18-27

[108] Driesse A, Thevenard D. A test of Suehrcke's sunshine-radiation relationship using a global data set. Solar Energy. 2002;**72**(2):167-175

[109] Muneer T, Zhang X. A new method for correcting shadow band diffuse irradiance data. Journal of Solar Energy Engineering-Transactions of the ASME Journal. 2002;**124**(1):34-43

[110] Yang K, Koike T. Estimating surface solar radiation from upper-air humidity. Solar Energy. 2002;**72**(2):177-186

[111] Chamkha AJ, Issa C, Khanafer K. Natural convection from an inclined plate embedded in a variable porosity porous medium due to solar radiation. International Journal of Thermal Sciences. 2002;**41**(1):73-81

[112] El-Adawi MK. New approach to modelling a flat plate collector: The Fourier transform technique. Renewable Energy. 2002;**26**(3):489-506

[113] Groenhout NK, Behnia M, Morrison GL. Experimental measurement of heat loss in an advanced solar collector. Experimental Thermal and Fluid Science. 2002;**26**(2-4):131-137

[114] Tripanagnostopoulos Y, Souliotis M, Nousia T. CPC type integrated collector storage systems. Solar Energy. 2002;**72**(4):327-350

[115] Rincon EA, Osorio FA. A new troughlike nonimaging solar concentrator. Journal of Solar Energy Engineering-Transactions of the ASME. 2002;**124**(1):51-54

[116] Belessiotis V, Mathioulakis E. Analytical approach of thermosyphon solar domestic hot water system performance. Solar Energy. 2002;**72**(4): 307-315

[117] Chen YT, Chong KK, Lim CS, Lim BH, Tan KK, Aliman O, et al. Report of the first prototype of non-imaging focusing heliostat and its application in high temperature solar furnace. Solar Energy. 2002;**72**(6):531-544

[118] Xi C, He-Ping L. Three-dimensional flow and heat transfer in thermal plasma systems. Surface and Coating Technology. 2003;**171**(1-3):124-133

[119] Ramachandran K, Nishiyama H. Three-dimensional effects of carrier gas and particle injections on the thermo-fluid fields of plasma jets. Journal of Physics D: Applied Physics. 2002;**35**(4):307-317

[120] Christlieb AJ, Hitchon WNG. Three-dimensional solutions of the Boltzmann equation: Heat transport at long mean free paths. Physical Review E. 2002:6708

[121] Zaman S, Mahwish G. Magnetohydrodynamic bioconvective flow of Williamson nanofluid containing gyrotactic microorganisms subjected to thermal radiation and Newtonian conditions. Journal of Theoretical Biology. 2019;**479**:22-28

[122] Sulochana C, Samrat SP, Sandeep N. Numerical investigation of magnetohydrodynamic (MHD) radiative flow over a rotating cone in the presence of Soret and chemical reaction. Propulsion and power Research. 2018;**7**(1):91-101

[123] Zhang J, Ming-Jiu N. Direct numerical simulations of incompressible multiphase magnetohydrodynamics with phase change. Journal of Computational Physics. 2018;**375**:717-746

[124] Kumaran G, Sandeep N, Animasaun IL. Computational modeling of magnetohydrodynamic non-Newtonian fluid flow past a paraboloid of revolution. Alexandria Engineering Journal. 2018;**57**(3):1859-1865

[125] Rahimah J, Roslinda N, Ioan P. Magnetohydrodynamic rotating flow and heat transfer of ferrofluid due to an exponentially permeable stretching/shrinking sheet. Journal of Magnetism and Magnetic Materials. 2018;**465**:365-374

Chapter 2

Applications of Compound Nanotechnology and Twisted Inserts for Enhanced Heat Transfer

Hussain H. Al-Kayiem, Muna S. Kassim and Saud T. Taher

Abstract

Nanoadditives are a type of heat transfer enhancement techniques adopted in heat exchangers to improve the performance of industrial plants through improvement of the thermal properties of base fluids. Recently, various types of inserts with nanofluids are adopted to enhance the thermal performance of double pipe heat exchangers. In the current article, TiO_2/water nanofluid with multiple twisted tape inserts was investigated as a hybrid enhancement technique of heat transfer in straight pipes. The investigations were carried out experimentally and numerically at Reynolds numbers varied from 5000 to 20,000. Using nanofluid with 0.1% TiO_2 nanoparticles volume fraction demonstrated enhanced heat transfer with slight increase in pressure drop. Results are showing a maximum increase of 110.8% in Nusselt number in a tube fitted with quintuple twisted tape inserts with 25.2% increase in the pressure drop. However, as the article is representing a part of specified book on heat exchangers, the literature has been extended to provide sufficient background to the reader on the use of nanotech, twisted inserts, and hybrid of compound nanofluids and inserts to enhance heat transfer processes.

Keywords: enhanced heat transfer, hybrid heat transfer enhancement, multi-longitudinal vortices, multiple twisted tapes, TiO_2 nanofluid

1. Introduction

Heat exchangers (HEXs) are typical thermal systems in industrial and engineering applications. They are adopted as means of heat dissipation in a wide range of thermal processes ranging from huge scale to microscale. HEXs are involved in the power production process, chemical and food industries, electronics cooling, environmental production engineering, waste heat recovery, manufacturing industry, air conditioning, and refrigeration applications. HEXs constitute a multibillion-dollar industry in the United States alone, and there are over 300 companies engaged in the manufacture of a wide array of heat exchangers [1].

The performance of heat exchanger can significantly increase by the heat transfer augmentation techniques that lead to the reduction of heat exchanger size, as well as operating cost reduction. Enhancement techniques can be classified either as passive or active techniques. A passive technique does not need any external power input, and the additional power needed to enhance heat transfer is taken from the available power in the system such as extended surfaces, treated surfaces, and twisted tape. Active techniques require external power such as mechanical aids and surface vibration [2].

Various methods of heat transfer enhancement are used in HEXs. These techniques are expressive to be manufactured and adopted to increase the thermal system efficiency by increasing the rate of heat transfer process and/or to reduce the size of thermal systems. They can be classified into two main categories; (i) active methods: which use an external power source; (ii) passive methods: which use several techniques without a power source such as turbulence generators (as propeller, spiral fin, twisted tapes, ribs ... etc.), or by using additives like the nanoadditives. However, in their review paper on the tape enhanced heat transfer [2], classified third enhancement techniques. They proposed compound methods, when two passive methods are used simultaneously. The method adopted in this paper is compound enhancement as twisted tapes and nanoadditives. As such, below are some backgrounds on the nanoadditives and twisted tapes to enhance heat transfer.

1.1 Nanoenhanced heat transfer

Nanoadditives are a type of heat transfer enhancement method through enhancement of the thermal properties of base materials. They may be synthesized with base fluids to produce nanofluids for thermal transport or synthesized with phase change materials (PCM) to produce nanocomposites for thermal energy storage as in [3].

Numerous works and review papers have been published in the field of nanoenhanced fluids, such as the paper of [4] who presented and compared the preparation, stability and thermophysical properties of nanofluids. It was concluded that nanofluids have, in general, better thermo-physical properties even at a very low particle concentration (typically 1.0% or less) than conventional heat transfer fluids. The only drawback is increment in the viscosity which leads to a higher pressure drop. Similar article by [5], published in 2018, has also reviewed the fabrication, stability, and thermophysical properties of nanofluids. After presenting the progress of studies on nanofluid thermophysical characterization, they identified some possible opportunities for future research that can bridge the gap between in-lab research and commercialization of nanofluids as this nanofluids are receiving large attention due to their potential usage. Reference [6], who reviewed the current scenario and future prospective of nanofluids, who also supported the conclusion of [5], ended with that the study of nanofluids has been materialized as a new field of scientific interest and innovative application. Reference [3] investigated nanofluid characteristics concerning thermal behavior in a plain tube, which showed significant enhancement rate in heat transfer with increase in nanoparticles diameter and volume concentration. Reference [7] experimented the thermophysical properties of Al_2O_3 and TiO_2 nanoparticles with distilled water and proposed parabolic equations for dynamic viscosity and thermal conductivity at specific conditions.

1.2 Enhanced heat transfer by twisted inserts

Twisted tape heat transfer enhancement mechanism is attributed to producing swirl flow in form of secondary recirculation act on the axial flow. This fluid mechanism increases the tangential and radial turbulent components leading to higher mix fluid layers and reduces the thermal and hydrodynamic boundary layer thickness due to increase in turbulent fluctuation and combination occurring between core region and near wall region in fluid flow, which cause higher temperature gradient near the wall leading to increased heat transfer rate.

Reference [8] reported for single twisted tape insert in circular tube in which heat transfer rate increased significantly than tube without inserts, and as twist ratio decreases, the heat transfer rate and friction factor become larger. Subsequently, [9] experimented the single twisted tape insert with many different twist ratios to describe the mechanism of heat transfer enhancement and proposed correlations for Nusselt number and friction factor.

A recognized contribution to the field of heat transfer enhancement is done by the thermofluid team in University of Pretoria—South Africa. They have started their investigation on heat transfer enhancement since 2010 and since then, they published a large number of experimental and numerical investigation on the enhanced heat transfer by inserts. In 2013, [10] investigated the heat transfer enhancement in laminar flow regime in circular tubes using rib roughening and twisted tape inserts. The work is carried out experimentally, and Nusselt number and the friction factor were measured. The work concluded major finding that the center-cleared twisted tapes in combination with transverse ribs perform significantly better than the individual enhancement technique acting alone for laminar flow through a circular duct up to a certain amount of center-clearance. The reported results are useful for the design of solar thermal heaters and heat exchangers. In 2017, another two papers have been published by the same team [11, 12]. Both papers have been carried out numerically. In [11], heat transfer behavior in a tube with inserted twisted tape swirl generator is investigated numerically, for different values of the twist ratio and diameter ratio and for Reynolds numbers within the range of 100–20,000. Results have shown that the tube use of twist tape enhances heat transfer generally, but, accompanied with a higher pressure drop. Improvement of the thermal-hydraulic performance can only be observed for certain configurations and Reynolds numbers. In [12], simulations were conducted for laminar, transitional, and turbulent flow regimes for four different rib angle of attack values and for a plain tube without ribs, as benchmark case. Within the investigated range, the larger thermal performance factors are observed to occur for the intermediate Reynolds numbers. Maximum values in the range of 2.0–2.5 are predicted for the Reynolds number of 2000, where a subsequent drop to values within the range of 1.0–1.5 is found to occur for Reynolds numbers around 3000–4000, which may be attributed to the transitional effects. In 2018, a paper [13] is published by the same team, which reported results of experimental investigation of heat transfer performance of corrugated tube with spring tape inserts in turbulent forced convection in Re range of 10,000–5000. Air is adopted as working fluid. Results show that Nusselt numbers can be increased considerably, depending on pitch and spring ratios. The heat transfer enhancement to the pressure drop penalty is realized to be larger than unity for all cases. Values around 2.8 occur for cases with the smallest pitch and spring ratios. Predictive Nusselt number and friction factor correlations are proposed.

1.3 Compound nanoadditive/twisted tape enhanced heat transfer

By combining the nanofluid enhancement with inserts, [14] found that using Al_2O_3 nanoparticles with distilled water and full-length twisted tape gave a superior performance with a noticeable increment in friction loses than plain tube. Reference [15], achieved 20 and 2.5% increase in overall heat transfer coefficient and friction factor, respectively, by using nanofluid with single twisted tape insert in double pipe heat exchanger than that without insert. Reference [16] presented an experimental analysis of the turbulent flow in tube fitted with (single, dual, triple, and quadruple) twisted tapes and nanofluid under turbulent flow conditions. The results shown that Nusselt number and friction factors increased as the number of tapes and volumetric concentration increased. Also, the increment in heat transfer rate by increase in nanofluid volumetric concentration only was higher than that of increase in twisted tape number only. It must be mentioned that the volumetric efficiency of the tube was not taken into consideration.

Reference [17] has carried out experimental and numerical investigations on similar tube flow and similar nanofluid with plain tube, tube fitted with dual plain tape inserts, tube fitted with dual twisted tapes inserts and tube fitted with dual helical screw twisted tape inserts. He concluded that, a maximum enhancement of 82.2% is achieved in the Nusselt number by using tube fitted with dual helical screw twisted tapes inserts and TiO_2/water nanofluid flow than that observed with the plain tube and distilled water flow. And the maximum friction factor observed for the same model of the tube fitted with dual helical screw twisted tape inserts and nanofluid are up to 17.34% than that of the plain tube. Reference [18] carried out a wide range of experimental study on the convective heat transfer enhancement using combined techniques. One of these techniques is the use of twisted tape along the whole tube length of a micro-fin tube that effectively combined the features of extended surfaces, turbulators, and artificial roughness. Nanofluids are used for improving the thermo-physical properties of the fluid. Ag-water nanofluid in a micro-fin tube with nonuniform twisted tapes insert is examined under turbulent flow. The effects of the twist ratios of nonuniform twisted-tapes of 3.0 > 2.8 > 2.6, 3.0 > 2.6 > 2.2, and 3.0 > 2.4 > 1.8, in counter and co-current flow arrangements and nanofluid concentrations of 0.007, 0.016, and 0.03% vol. are investigated. They claimed that heat transfer, friction loss, and thermal performance factors are increasing as the twist ratios are decreasing for nonuniform twisted tapes and increasing nanofluid concentrations. The optimum condition are achieved in using the micro-fin tube with a nonuniform twisted-tape in a counter-current-arrangement with twist ratios in a series of 3.0 > 2.4 > 1.8 with Ag-water nanofluid at a concentration of 0.3% vol. The enhancements are up to 112.5% for the heat transfer rate and 1.62 for thermal performance.

On other combined enhancement attempt in microchannel heat exchangers, [19] carried out experimental investigation on heat transfer for pulsating flow of GOP-water nanofluid. The effects of mass fraction of graphene oxide (GOPs) and flow pulsating frequency on heat transfer and pressure drop in a microchannel with arrayed pin-fins have been investigated. Five different mass fractions of graphene oxide nanofluids were prepared and used as working fluids. Experiments were performed under the condition that the pulsating frequency was from 1 to 5 Hz, the mass fraction was from 0.02 to 0.2%, and the average Reynolds numbers were 272, 407, and 544. The results show that the heat transfer is enhanced significantly when the frequency is in the range of 2–5 Hz. For the frequency of 1 Hz, the pulsating flow shows a negative effect on temperature uniformity. With the increase of mass fraction, the heat transfer performance is improved, while no significant change is found in pressure drop. The pulsating flow leads to a significant enhancement of pressure drop for frequency at 2 Hz. The combination of pulsating and nanofluid can obtain higher heat transfer efficiency under limited size of microchannel heat sink and low inlet Reynolds numbers. The results provided good guide for the design of microchannel heat exchangers.

Since conventional fluids, such as water, have a relatively poor heat transfer characteristic, the nanoenhancing technique opens the door to gain more benefits from these conventional fluids especially in heat transfer intensification field.

As pointed above and mentioned by many other researchers [20–23], the field of nanoenhanced heat transfer with multiple inserts is still virgin. Further experimental data are essential to support the literature, and to enhance the understanding of hydrothermal behavior in thermal systems. So far, few studies have been carried out on multiple twisted tapes with nanofluid effect on thermo-hydraulic characteristics in thermal systems.

The objective of this chapter is to scrutinize the effect of compound multiple twisted tape (TT) inserts with TiO_2-water nanofluid on heat exchange enhancement in double pipe heat exchangers. Three cases of tubes fitted with single, triple, and quintuple plain twisted tapes have been investigated experimentally and simulated numerically. In addition, two bench-mark cases have been investigated: first case is with plain tube with pure water flow, and the second case is with plain tube with nanofluid flow. Results have been manipulated and presented in terms of Nusselt number for heat transfer and friction factor for pressure drop.

2. Problem formulation

The basic geometry adopted in this investigation is a straight tube with 1000 mm length, L and 50 mm internal diameter, D. Pure tube, with pure water flow and no inserts, was considered as the benchmark case to compare the thermal enhancement and the pressure drop. The other cases were investigated with a different number of inserts and with 0.1 vol.% TiO_2/water nanofluid.

It is worth mentioning that this research meant to investigate the flow characteristics only in compound thermal and hydrodynamic in fully developed region where the entrance effect becomes insignificant beyond a pipe length of 8 times the diameter for turbulent flow [8, 14]. The fully developed region was calculated to be accomplished at 500 mm from the tube inlet.

2.1 Twisted tape inserts

Twisted tape inserts (TT) are heat transfer enhancement devices which are dividing the flow within the tube resulting in higher velocity near the tube surface. They, also, creating spiral flow creates swirl or secondary flow in the main flow which increases local velocities and promotes mixing. They are widely used over decades to generate the swirl flow in the thermal fluid resulting in increased heat transfer coefficient, with a penalty of increased pressure drop across the flow passage. Thus, reduction in the thermal system, like the heat exchangers, can be achieved. Types of TT are shown in **Figure 1a**.

Main parameters that are commonly adopted to characterize the TT are the empty tube Reynolds number (Re), half-pitch (y), and twist ratio (Y). The main geometrical TT identifiers are shown in **Figure 1b**.

The half-pitch (y) is defined as the distance between two points on the edge of a TT, which lies down on the same plane as the TT completes 180° of revolution.

The twist ratio, Y, is defined as the ratio of the half-pitch to the internal tube diameter, $Y = y/D_p$.

Tube fitted with single twisted tape (STT), tube fitted with triple twisted tapes (TTT), and tube fitted with quintuple twisted tapes (QTT) are considered in the present experimental and numerical investigations. The schematics of cross sectional view of these models and twisted tapes and the geometries are illustrated in **Figure 2**.

The twisted tapes have the same length of the tube and has a width; L = 1000 mm, width, w (mm), thickness, δ (mm) and pitch of 180° twist, y (mm) as explained in **Table 1**. The swirl direction corresponding to tape arrangement was co-swirl flow, and all tapes were aligned to be twisted in the same direction.

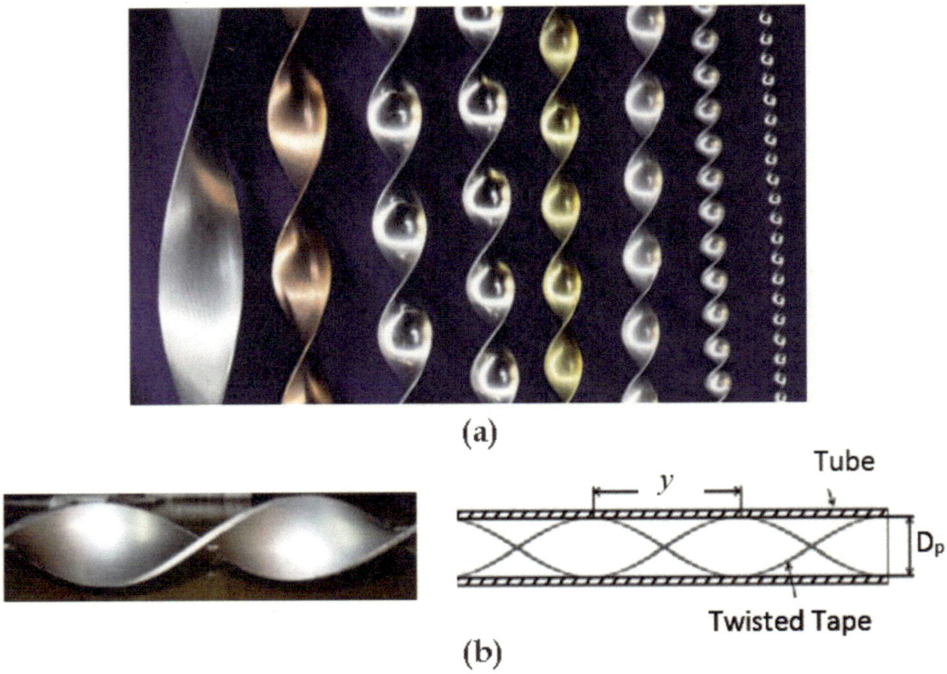

Figure 1.
Twisted tape inserts; (a) different types of twisted tape inserts (courtesy of visual capitalist [24], with permission); (b) topologies of twisted tape insert.

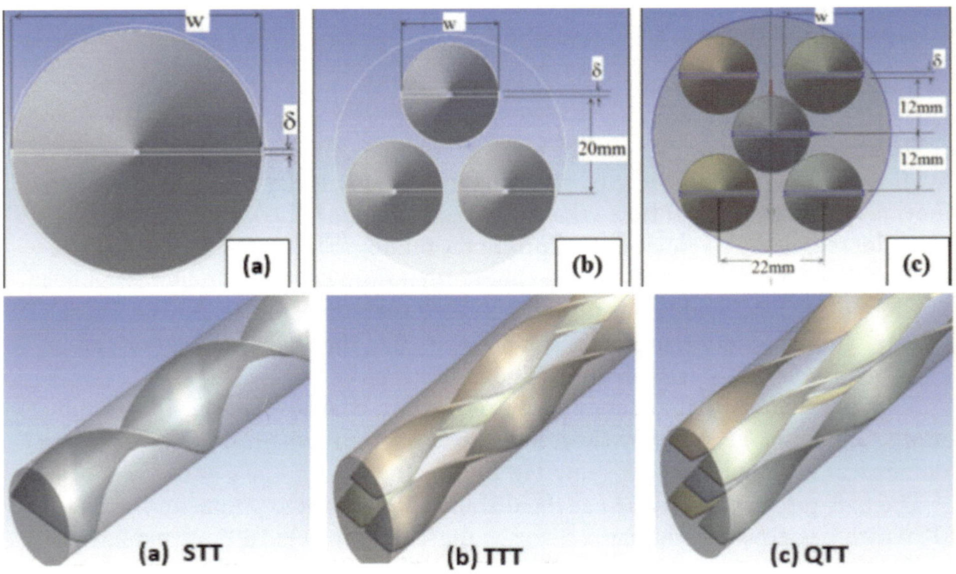

Figure 2.
Schematic views for tubes fitted with; (a) single twisted tape (STT), (b) triple twisted tapes (TTT), and (c) quintuple twisted tapes (QTT).

2.2 Working fluids

Water was used in the primary tests to investigate the effect of inserts only on the thermo-hydraulic performance of tubes. Latterly, nanofluid of TiO_2 pre-dispersed in water was used to obtain the enhancement by both nanotechnology

Model	No. of tapes	Tape width (mm)	Tape pitch (y) (mm)	Tape length (mm)	Tape thickness (δ) (mm)	Tape material
STT	1	48	100	1000	1	Aluminum
TTT	3	21	100	1000	1	Aluminum
QTT	5	16	100	1000	1	Aluminum

Table 1.
Geometries and details of TT inserts.

	Density ρ (kg/m^3)	Specific heat (C_p) (J/kg·K)	Thermal conductivity, k (W/m·K)	Dynamic viscosity, μ (kg/m·s)
TiO_2	4230	692	8.4	—
Water	997	4179	0.613	855×10^{-6}

Table 2.
Properties of titanium oxide particles and water.

Property	Formula	Relation no.	Recommended by	Value
Density, ρ	$\rho_{nf} = \varphi\rho_{np} + (1-\varphi)\rho_{bf}$	(1)	Pak and Cho [25]	1000.2 kg/m^3
Specific heat, C_p	$(C_P)_{nf} = \varphi(C_P)_{np} + (1-\varphi)(C_P)_{bf}$	(2)	Maddah et al. [15]	4175.5 J/kg·K
Thermal conductivity, k	$k_{nf} = [1+3\varphi])k_{bf}$	(3)	Said et al. [7]	0.6148 W/m·K
Dynamic viscosity, μ	$\mu_{nf} = [1+2.5\,\varphi + 6.2\,\varphi^2])\mu_{bf}$	(4)	Batchelor [26]	857.143×10^{-6} kg/m·s

Note: φ is the percentage of volume fraction of nanoparticles.

Table 3.
Correlations of 0.1% volume of TiO_2-in-water nanofluid properties and predicted values.

and twisted tape inserts. In the present study, 0.1 vol.% TiO_2 nanoparticles of size (<50 nm) mixed with distilled water was sonicated continuously by ultrasonic vibrator generating pulses of 240 W at 40 + 4 kHz to break down any possible nanoparticle agglomeration. Nanofluid was found to be stable during tests period of tests and no intermediate mixing process considered necessary. TiO_2 particles and water have the considered properties illustrated in **Table 2**.

Volume concentration φ desired in this study was 0.1% for TiO_2/water. The thermophysical parameters are calculated for nanofluid; **Table 3** contains the popular and valid models, which were used to evaluate property values and shows values.

3. Experimental implementation

3.1 Apparatus

Experimental setup was designed and fabricated to investigate the thermo-hydraulic characteristics of the double pipe heat exchanger, as shown in **Figure 3**. The test section comprised of copper tube of 1000 mm length with inside and

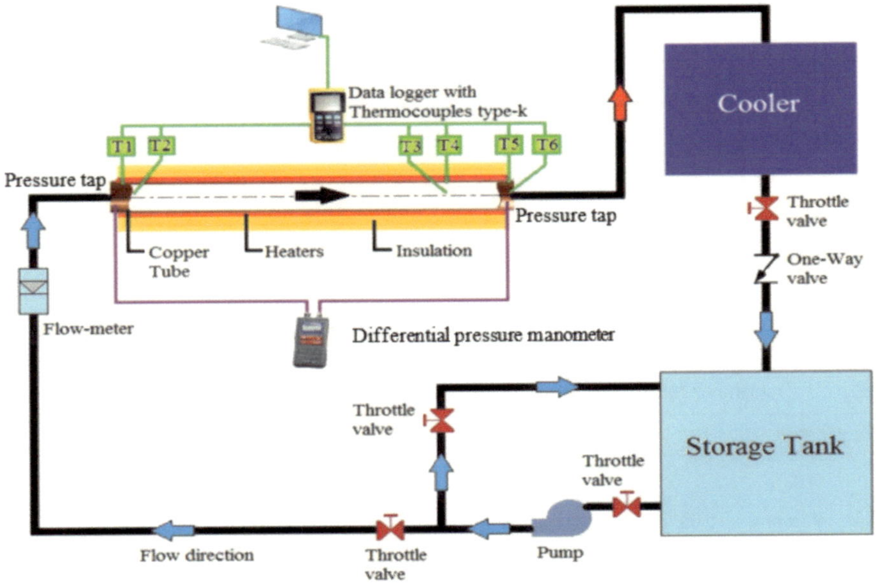

Figure 3.
Schematic diagram of the experimental setup.

outside diameters of 50 and 53 mm, respectively. An electric heater with 5 kW capacity was wounded on the outer surface of the copper tube and insulated by glass-wool insulation. Thermocouples, referred to as (T1–6) in **Figure 3**, were installed to measure wall and fluid temperature at inlet, fully developed, and outlet regions. All thermocouples were calibrated before being fixed, and all were connected to the data logger to record temperature readings. A differential pressure manometer was used to measure pressure drop along the test section. The working fluid was heated inside the test section and allowed to cool by passing through a cooler (evaporative cooling system). By recirculation, the working fluid returns to the storage tank in the flow loop and then is pumped again to the test section. Flow meter was placed at the entrance of the test section to measure the working fluid flow rate. Throttle valves were incorporated to allow controlling the working fluid flow rate and for maintenance and emergency, if any. To ensure steady state condition for each run, period of 30 minutes was permitted prior to starting the measurement and data acquisition.

3.2 Verification of experimental measurements

Heat transfer in convective flow is commonly evaluated in terms of Nusselt number, Nu, as in Eq. (5).

$$\mathrm{Nu} = \frac{h.D_p}{k} = \frac{\left[\bar{\bar{q}} / (T_s - T_b)\right] D_p}{k} \tag{5}$$

Nusselt number values were validated with the experimental results of previous researches shown in **Table 4**. Values of Reynolds numbers, Re, were evaluated as per Eq. (6), as follows:

$$\mathrm{Re} = \frac{VD_p}{v} \tag{6}$$

where v is the mean velocity (=measured flow rate/pipe inlet cross section area), ν is the kinematic viscosity of the working fluid, and D_p is the pipe or tube diameter.

Prandtl number is considered 7.56 for water. For nanofluid, Pr is evaluated using Eq. (7), as follows:

$$\mathrm{Pr} = \frac{Cp_{nf} \times \mu_{nf}}{k_{nf}} \tag{7}$$

The friction factor was predicted using the known Darcy-Weisbach equation (Eq. 11), for both, water and nanofluid flow, as follows:

$$f = \frac{\Delta p}{0.5\, \rho\, V_{in}{}^2 \left(\frac{L}{D_h}\right)} \tag{11}$$

where Δp is the measured pressure drop over the 1.0 m pipe length. The measured values were vindicated with results gained from well-established correlations, illustrated in **Table 5**.

Correlation	Eq. No.	Suggested by	Case of HEX
$Nu = \frac{\left(\frac{f}{8}\right)\left(Re_{D_h} - 1000\right)Pr}{1+12.7\left(\frac{f}{8}\right)^{\frac{1}{2}}\left(Pr^{\frac{2}{3}}-1\right)}$	(8)	Gnielinski, in [27]	Tube with water flow
$Nu = 0.074\, Re_{nf}^{0.707} Pr_{nf}^{0.385} \varphi^{0.074}$	(9)	Duangthongsuk and Wongwises [21]	Tube with nanofluid flow
$Nu = 0.024\ Re^{0.8} pr^{0.4} \times \left[1 + \frac{0.769}{\frac{y}{w}}\right]^{1.1} \left[\frac{T_b}{T_s}\right]^{0.45} \left[\frac{\pi+2-\frac{2\delta}{D}}{\pi-\frac{4\delta}{D}}\right]^{0.2} \left[\frac{\pi}{\pi-\frac{4\delta}{D}}\right]^{0.8}$	(10)	Bergles and Manglik [9]	Tube with twisted tape(s) and water flow

Table 4.
Adopted correlations from the literature to predict Nusselt number used to verify the experimental results.

Correlation	Eq. No.	Suggested by	Case of HEX
$f = 0.318\, Re_{D_h}{}^{-0.25}$	(12)	Blasius, in [27]	Tube with water flow
$f = \left(0.79 \ln Re_{D_h} - 1.64\right)^{-2}$	(13)	Petukhov, in [27]	Tube with water flow
$f = 0.961 \varphi^{0.052} Re_{D_h}{}^{-0.375}$	(14)	Duangthongsuk and Wongwises [21]	Tube with nanofluid flow
$f = \left[0.079\, Re^{-0.25} \left[\frac{\pi+2-\frac{2\delta}{D}}{\pi-\frac{4\delta}{D}}\right]^{1.25} \left[\frac{\pi}{\pi-\frac{4\delta}{D}}\right]^{1.75}\right] \left[1 + 2.06\left(1 + \left(\frac{2\frac{y}{w}}{\pi}\right)^2\right)^{-0.74}\right]$	(15)	Bergles and Manglik [9]	Tube with triple twisted tapes and water flow

Table 5.
Adopted correlations from the literature to predict friction factor used to verify the experimental results.

3.3 Experimental uncertainty and errors

The used instruments were well calibrated, and the margin of error and uncertainty of each instrument were evaluated.

- Thermocouple type-K used in this study to measure temperature of surfaces. Thermocouple type-K probes were used to measure the fluid temperature. All thermocouples and probes were calibrated and certified by the Central Organization for Standardization and Quality Control (COSQC)—Baghdad—Iraq [certificate number 734/2016 dated 30 Oct. 2016]. Mean uncertainty within the temperature measurement in the experiment was $\pm 0.63°C$.

- Flow meters type LZS-25 used to measure the flow rate of heat transfer fluids were calibrated using a scaled container and time recording technique, which is well identified as a standard procedure for flow meter calibration. The relative uncertainty of flow measurement was $\pm 2.34\%$.

- Differential manometer was calibrated against another certified manometer type Lutron (PM-9100). The relative uncertainty was $\pm 2.1\%$.

- The two pressure gauges were calibrated and certified by the Central Organization for Standardization and Quality Control (COSQC)—Baghdad—Iraq [certificate number PRE/918/2016 dated 06 Oct. 2016]. Max uncertainty was ± 0.012 bar.

The possible error in the prediction of Re and Nu was estimated. The uncertainty in Re values was estimated by Eqs. (16)–(18) using the uncertainties of the measuring instruments used for measurements of relevant variables in Re:

$$Re = \frac{4\,\dot{m}}{\pi \mu d} = \frac{4 \rho \dot{V}}{\pi \mu d} \tag{16}$$

$$\Delta\, Re = \left[\left[\left(\frac{\sigma Re}{\sigma \dot{V}} \right) \bullet \dot{V} \right]^2 + \left[\left(\frac{\sigma Re}{\sigma d} \right) \bullet d \right]^2 \right]^{0.5} \tag{17}$$

Getting,

$$\Delta\, Re \,/\, Re = \left[\left[\left(\frac{\Delta \dot{V}}{\dot{V}} \right) \right]^2 + \left[\left(\frac{\Delta d}{d} \right) \right]^2 \right]^{0.5} \tag{18}$$

= 0.0665 or 6.65%.

Uncertainty in Nu values was estimated by Eq. (19) using the uncertainties of the measuring instruments used for measurements of relevant variables in Nu, as

$$\Delta Nu / Nu = \left[\left[\left(\frac{\Delta h}{h} \right) \right]^2 + \left[\left(\frac{\Delta d}{d} \right) \right]^2 \right]^{0.5} \tag{19}$$

= 0.0625 or 6.25%

4. Computational methods

4.1 Numerical simulation and conditions

The numerical analyses included single phase flow based on steady and three dimensional continuity; momentum and energy equations were performed using the ANSYS-Fluent commercial code. RNG k-ε model was chosen to solve the considered cases because the effect of swirl on turbulence is included in this model, which enhances the accuracy of swirling flows.

4.2 Boundary conditions

The thermo-fluid process in the current simulated was solved as steady, incompressible, and 3D flow. The following was adopted to set the boundary conditions for the simulation:

- The flow was considered as internal flow under uniform heat flux condition for all cases with $\overline{\overline{q}}$ = 30,000 W/m^2.
- No slip condition was also applied to the tube wall.
- Results are considered only at the fully developed region ($L > 10D_{\mathrm{p}}$).
- Water and 0.1 vol.% TiO_2/water are considered as the working fluids, and calculations for the thermal properties are done at the inlet temperature of 300 K, as in **Tables 2** and **3**.
- The nanofluid was considered as a single-phase fluid with changed physical parameters as density, thermal capability, thermal conductivity, and viscosity.
- All cases were investigated within the turbulent flow regime studied with a range for Reynolds number of 5000–20,000.
- The hydraulic diameter at each inlet and outlet has the same value.
- At the outlet, a pressure outlet condition was used, and the gauge pressure is set to zero.
- Turbulence intensity obtained from the expression recommended by [28].

$$T.I. = 0.16\,Re_{D_p}^{-0.125} \tag{20}$$

Other flow quantities are extrapolated from the interior domain by the solver in Fluent software. The SIMPLE (Semi-Implicit Pressure Linked Equations) algorithm were chosen as solver method. In addition, a convergence criterion of 10^{-5} was used for energy and mass conservation of the calculated parameters.

4.3 Grid generation

To avoid blurred curved areas, tetrahedral cells are used for meshing the computational domains, as shown in **Figure 4**. Mesh, with around 3,000,000 elements, was decided to represent the domains for the current simulation.

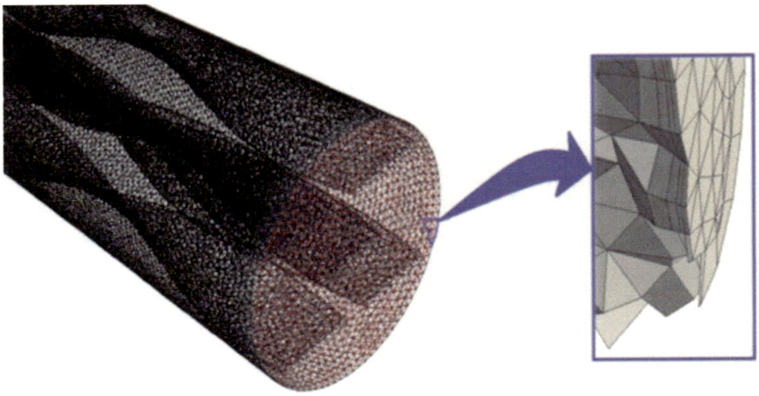

Figure 4.
Tetrahedral mesh cells for the case of QTT with enlarged small cut for clarification.

Orthogonal Quality mesh metrics					
Unacceptable	Bad	Acceptable	Good	Very good	Excellent
0-0.001	0.001-0.14	0.15-0.20	0.20-0.69	0.70-0.95	0.95-1.00

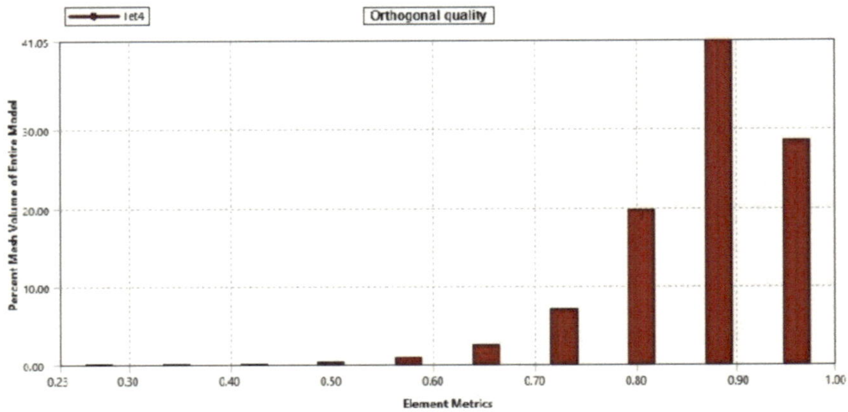

Figure 5.
Mesh cell orthogonal quality metrics.

Skewness mesh metrics spectrum					
Excellent	Very good	Good	Acceptable	Bad	Unacceptable
0-0.25	0.25-0.50	0.50-0.80	0.80-0.94	0.95-0.97	0.98-1.00

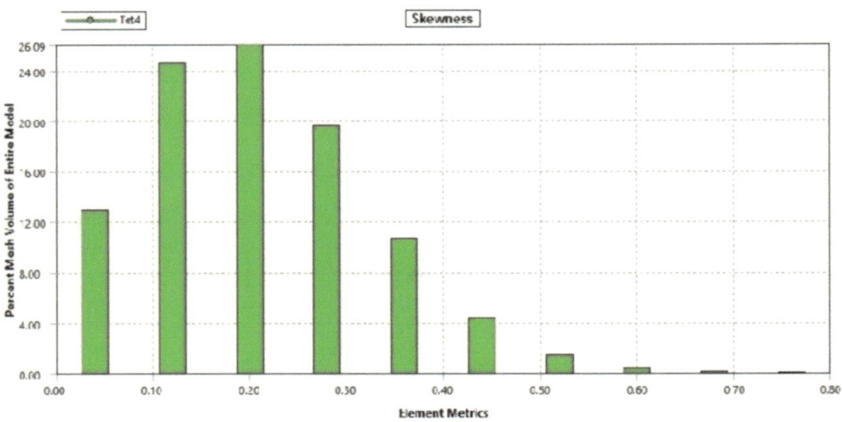

Figure 6.
Mesh cell skewness quality metrics.

The same way that element numbers in the mesh are important, mesh quality has a remarkable role in the numerical solution accuracy. Grid quality is commonly identified through orthogonal quality and skewness. Orthogonal quality describes how much the mesh criteria are within the correct range that is valid for physical value prediction. Orthogonal quality is presented in **Figure 5** for the grids performed to represent computational domains. The average orthogonal quality value gained in the present numerical procedure, of 0.8563, is within a very good quality range. Skewness determines how the generated cells are close to the ideal configuration and it governs solution ability to converge, as illustrated in **Figure 6**. The average skewness value of depended grids of 0.22288 is within excellent simulation range.

5. Results and discussion

The presentation and discussion of results are not based on sub-sectioning the numerical and experimental results. Instead, the results are sub-sectioned based on the hydrothermal parameters, like the velocity, pressure, and thermal performance.

5.1 Hydrodynamics analysis

The hydrodynamics analysis of the flow within internal conduits includes the velocity and pressure structures.

5.1.1 Velocity field analysis

The flow structure in the pipe is characterized by analyzing the velocity field and the pressure distribution in the flow domain. In the current work, there are four different geometrical configurations including plane pipe flow and another three cases with single, triple, and quintuple twisted tape inserts inside the pipe. CFD is a powerful tool to provide flow visualization and assist in the analysis of the flow field structure. Fields of velocity, predicted by computational simulation, in case of water flowing are depicted in **Figure 7**. The velocity contours shown are taken for

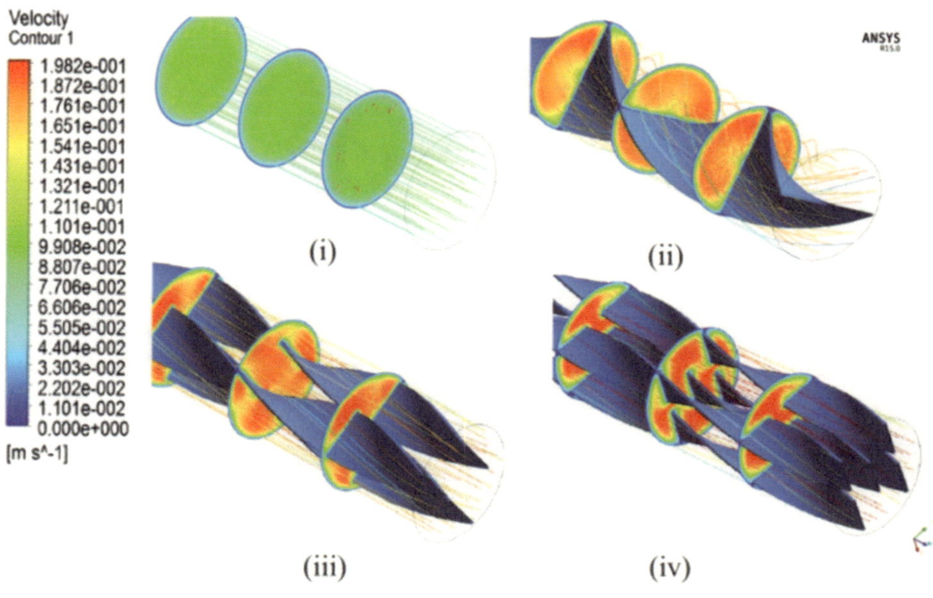

Figure 7.
Velocity contours and streamlines of water flow for; (i) PT, (ii) STT, (iii) TTT, and (iv) QTT.

Re = 5000 and at different axial locations in the fully developed region. As can be seen in **Figure 7**, frame ii, the velocity is increasing by 50.0% for STT higher than PT. This happens due to secondary flow caused by twisted tape geometry, which increases mean velocity by changing flow type from linear motion to swirl motion and also due to reduction in hydraulic diameter which leads to increase in velocity at constant Reynolds number and the reduction in hydraulic diameter causing a decrease in flow cross sectional area, which resulted in an increment in mean velocity value to satisfy equation of continuity (the rate of mass enters a system is equal to the rate of mass leaves the system).

For TTT and QTT, the velocity increased by 13.3 and 27.4% higher than STT due to narrowing the flow passages. Swirl motion and turbulence fluctuation are also increasing by increases in tape number due to multi-passage flow interactions. In addition, the tapes are breaking the flow field uniformity, and mix fluid flow layers between near wall region and core region lead to the appearance of many regions of high velocity; high velocity region in plain tube appears only at the core of it, which increases as the number of twisted tape increases, and that fission leads to increase in average velocity of fluid flow.

Longitudinal vortices in flow fields are shown in **Figure 8**; it was found that the number of vortices generated in the flow equals the number of twisted tapes and formed around it.

Apparently, the velocities of the nanofluid are nearly the same as those of water under the considerable nanoparticle volume fraction, which discloses that nanofluid will not require an added disadvantage over pumping power.

5.1.2 Analysis of pressure field and pressure drop

As a fluid flows through the tube, there will be a pressure drop due to the shear drag at the contact wall in addition to the pressure required to pump the fluid inside the tube which is in tube with inserts higher than those without. The main determinants of pressure drop are fluid viscosity and fluid velocity. Pressure contours are illustrated in **Figure 9** for the computational domains considered at Reynolds number 5000 for water flow and on longitudinal revolution surface along the axial direction.

In the case of STT, **Figure 9**, frame a, pressure drop is 98.7% higher than in PT due to the fact that twisted tape insert increases frictional shear forces within the surface area of the inserts.

Figure 10 presents the predicted pressure drop for all simulated cases including pure tube and tube with single, triple, and quintuple twisted tape inserts with water

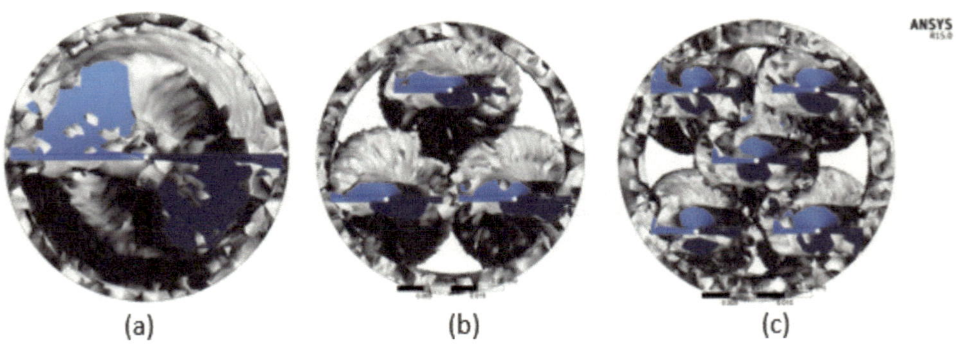

Figure 8.
Longitudinal water vortices for tube fitted with; (a) single twisted tape, (b) triple twisted tapes, and (c) quintuple twisted tapes.

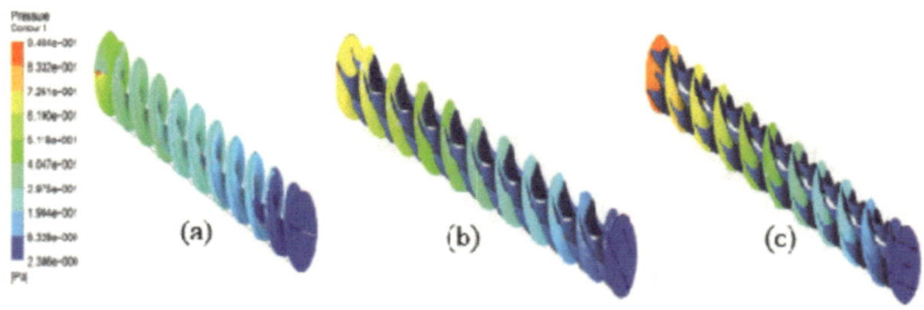

Figure 9.
Pressure contours for a longitudinal revolution surface along the computational domains of water flow; (a) STT, (b) TTT, and (c) QTT.

flow and with nanofluid flow. The pressure drop in TTT and QTT is 31.2 and 64.5% higher than STT, respectively. This is due to swirl motion achieved by each one of them, where the secondary motion generated by twisted tapes have an effect on velocity proportionally, where velocity gradient effect on shear forces acting on fluid flow causes pressure drop.

Pressure drop increases slightly by using nanofluid at 0.1 vol.% TiO_2. The numerical results show a percentage difference between water and nanofluid up to 3% for the same model investigated. However, experimental results for pressure agree well with those calculated numerically with a maximum divergence of 6.8%. It is obvious, from **Figure 10**, that pressure drop of water and for nanofluid increases with increasing Reynolds number. The small increase in pressure drop of nanofluid than water illustrates that using nanofluids with higher particle volume fraction may cause small penalty in pressure drop.

5.1.3 Friction factor

The friction factor is influenced by velocity variation, pressure drop, and contact surface topologies with the fluid flow. The models were examined numerically and experimentally in Reynolds number range varying from 5000 to 20,000.

The experimental results of the friction factor are reasonably matching with the results obtained from correlations 12 and 13 for the plain tube with water flow with

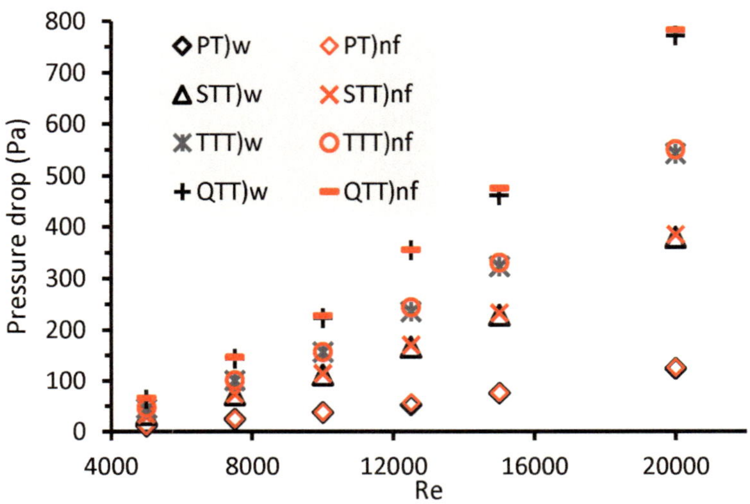

Figure 10.
Simulation results of pressure drop obtained from water (w) and 0.1 vol.% TiO_2/water nanofluid (nf) for different inserts at various Reynolds numbers.

a maximum deviation of 4.1%. Further verification was carried out by comparing the experimental results of the friction factor of single TT insert in the tube with water flow by comparison with the results gained by a correlation developed by Bergles, as in Eq. (15). The maximum deviation was 3.0%. The verification results are shown in **Figure 11**.

Measured friction factors coincide well with the calculated values from correlations of validation. As the fluid velocity increases, the friction factor decreases. Therefore, friction factor decreases with Reynolds number increasing. This is because Reynolds number increases the momentum, overcomes the viscous force of the fluid, and consequently lowers the shear between the fluid and the tube wall.

Reference [16] developed a correlation for the friction factor prediction for nanofluid flow in a plain tube, shown as Eq. (14). Comparison between the predicted results by Eq. (14) and the experimental results in the current investigations are shown in **Figure 12**. Good match between the experimental and empirical

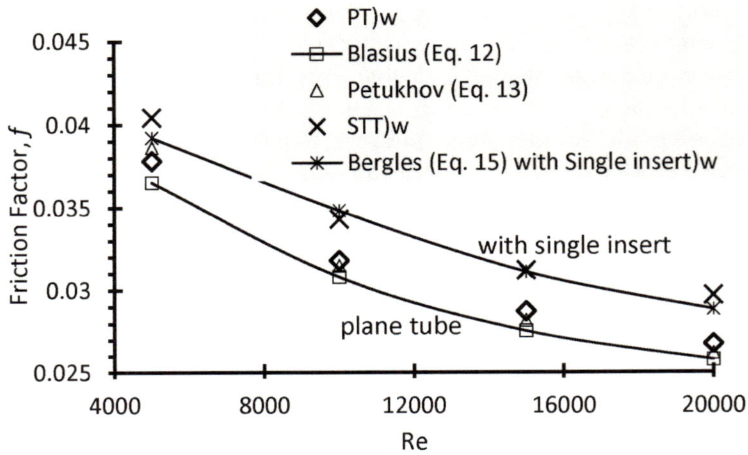

Figure 11.
Verification of experimentally measured friction factors and those predicted from; (a) PT with water flow and prediction by Blasius (Eq. 12) and Petukhov (Eq. 13); (b) STT with water flow and prediction by Bergles (Eq. 15).

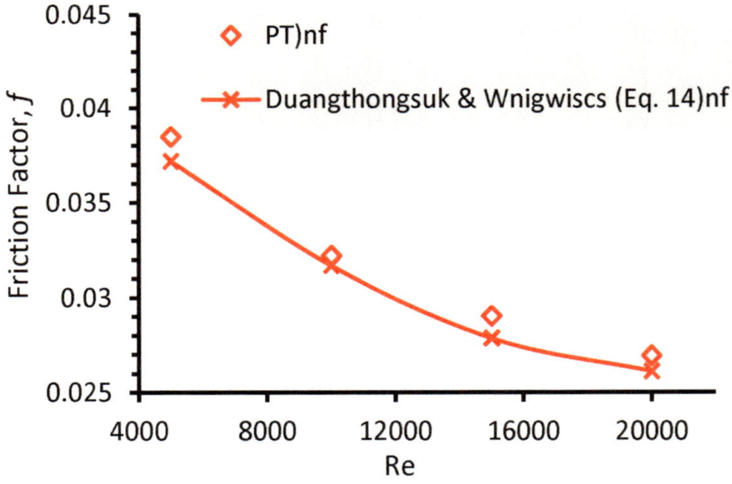

Figure 12.
Verification of experimentally measured friction factors and predicted by Duangthongsuk and Wongwises (Eq. 14) for 0.1 vol.% TiO_2/water nanofluid flow in plain tube.

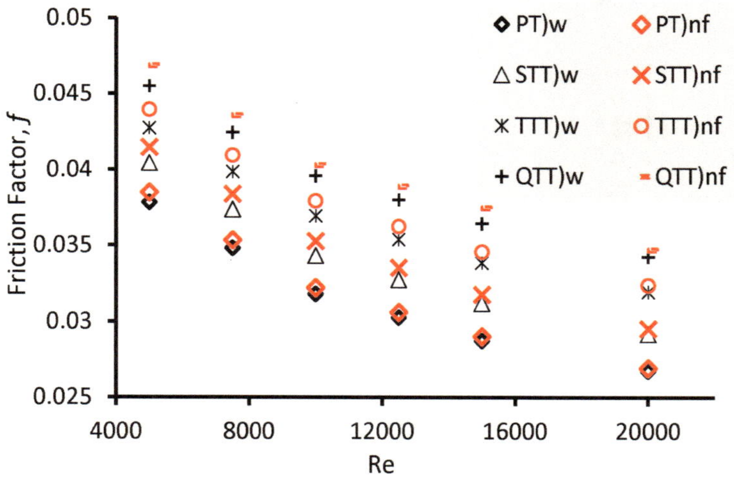

Figure 13.
Experimental measurement results for the friction factor verses Reynolds number for plain tube and tubes with single, triple, and quintuple twisted tape inserts of flow cases of water flow and 0.1 vol.% TiO_2/water nanofluid flow.

values has been achieved. Within the tested range of 5000–20,000 Re, the correlation shows overprediction of around 2.0–3.0% in the friction factor values compared to the experimental results.

The variations of the friction factor with Reynolds number for the tubes with different twisted tape inserts, with flows of water and nanofluid, are compared in **Figure 13**. The friction factors of the nanofluids are slightly higher than those of the base liquid. The tube fitted with plain twisted tape inserts (STT) when water flow has friction factor of 6.6–8.7% higher than plain tube. This is attributed to the flow blockage and swirl flow due to tape insert; for the same case, friction factor increases by 1.2–2.4% using nanofluid.

The additional dissipation of pressure of the fluid caused by the fluid disturbance due to increase of tape number results in an increase in pressure drop, which causes increase of friction factor. As the number of inserts increases, the pressure drop significantly increases. For water flow, the friction factors of TTT and QTT are 12.2–17.74% and 18.428–24.65% higher than that in plain tube. For nanofluid flow, the friction factors of TTT are 13.3–18.4% and for QTT are 19.6–25.2% higher than that in plain tube.

Results of predicted friction factor from the numerical simulation show the same trend as of the experimental ones, where differentiations between results within 6.04% are considered as an acceptable limit. Numerical simulation results of the friction factors for cases of inserts using water and nanofluid are shown in **Figure 14**. All cases demonstrated a slight increase in the numerically predicted friction factor values when 0.1 vol.% TiO_2 is used as working fluid. This is attributed to the slight increase in the viscosity of the nanofluid compared to the viscosity of pure water.

5.2 Thermal analysis

The heat transfer enhancement, in terms of Nusselt number, is influenced by velocity variation, friction factor, nanoparticles volume fraction, twisted tape dimensions, and other parameters. The four models were examined experimentally and numerically within Reynolds number ranging from 5000 to 20,000. The

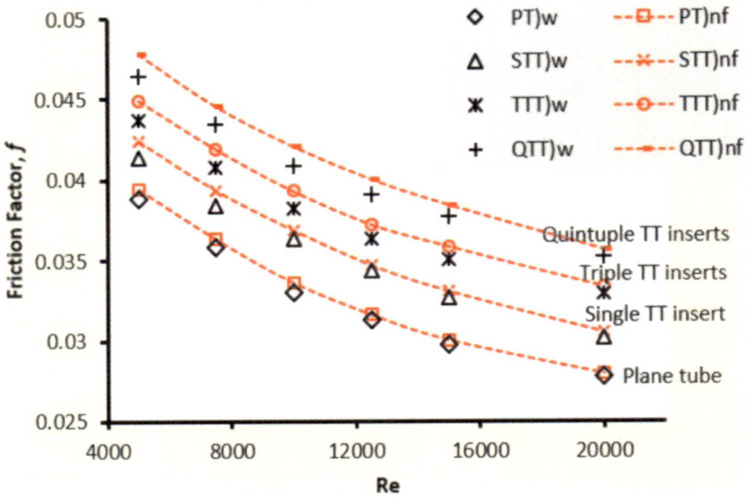

Figure 14.
Numerical simulation results of the friction factor versus Reynolds number for plain tube and tubes with single, triple, and quintuple twisted tape inserts of flow cases of water flow and 0.1 vol.% TiO_2/water nanofluid flow.

measured thermal parameters were verified by comparing the measurement results with well-established correlations to predict Nu. The verification results are shown in **Figure 15a** and **b**. The experimental results are matching those results obtained from correlations 8 and 15 with a deviation of 2.6–7.4%.

Further verification was carried out for the case of nanofluid flow in the pipe by comparing the experimental measurement Nu results with correlation 9 prediction Nu results, as shown in **Figure 16**. Very good agreement between the experimental and correlation results was demonstrated. The predicted results of Nu by the correlation are higher than the experimental results of Nu. As Re increased, a slight increase in the margin of error was observed.

The variations of Nusselt number with Reynolds number for the tubes with different twisted tape inserts are compared in **Figure 17** for both nanofluid and base fluid. It can generally be observed that the Nusselt number increases as the Reynolds number increases. This arises as a result of the momentum that overcomes the viscous force of the fluid as the Reynolds number increases and in effect diminishes the shear between the fluid and the tube wall.

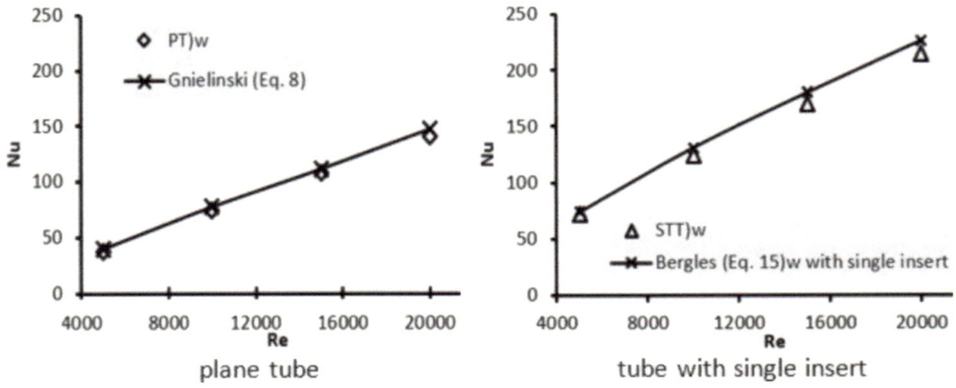

Figure 15.
Verification of experimental results for heat exchange performance by: (a) comparison with Gnielinski correlation for water flow in plain tube. (b) Comparison with Bergles correlation for water flow in tube with single TT insert.

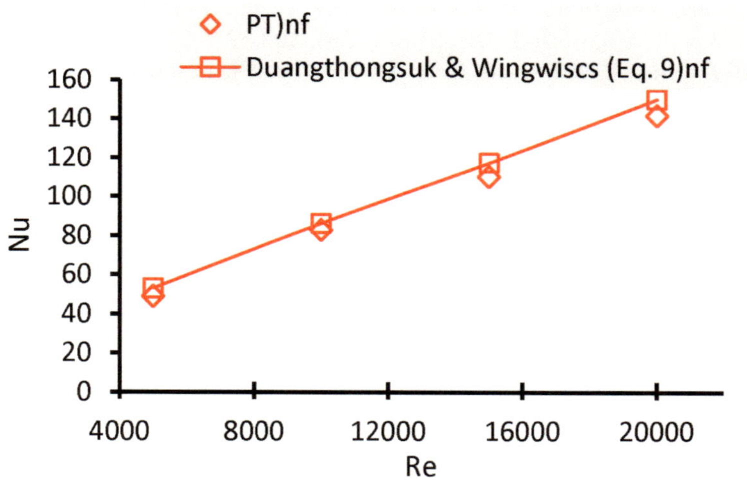

Figure 16.
Validation of experimental results for Nusselt number to the plain tube with Duangthongsuk and Wingwiscs correlation for 0.1 vol.% TiO_2/water nanofluid flow.

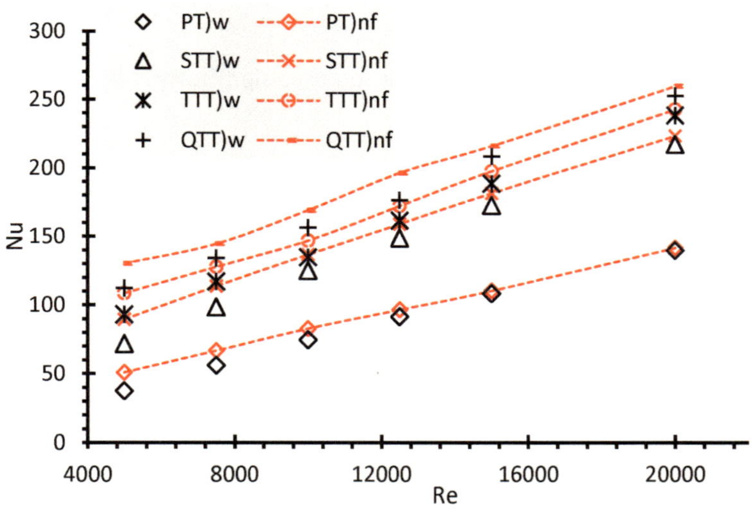

Figure 17.
Variation of Nusselt number with Reynolds number for plain tubes and tubes with single, triple, and quintuple twisted tape inserts operating with water flow and 0.1 vol.% TiO_2/water nanofluid flow.

For water flow, the STT has Nusselt numbers of 43.4–63.2% higher than plain tube. This enhancement in heat transfer rate returns to the act of twisted tape that generates swirl motion, which leads to the better temperature distribution at the core region and increases turbulence intensity at near wall region that results in higher temperature gradient there and enhances heat transfer coefficient. Also, secondary flow with greater enhancement was realized at higher Reynolds numbers.

In case of water flow, Nusselt numbers of TTT and QTT are 50.12–83.74% and 57.48–100.066% higher than that in measured in PT. It is obvious that Nusselt number increases with increasing twisted tape number due to increase in secondary motion violence, which disperses the high temperature region near wall to uniformly distribute all over flow cross sectional area, which is recognized as the key factor of heat transfer enhancement.

For 0.1 vol.% TiO_2/water nanofluid flow, PT, STT, TTT, and QTT have Nusselt numbers of 1.3–30.4%, 46.1–83.2%, 53.8–97.5%, and 59.9–110.8%, respectively, higher than that in PT with water flow. This behavior is due to the fact that nanoparticles presented in the base liquid increase the thermal conductivity, which leads to an increase in heat transfer performance.

6. Conclusions

Heat transfer and fluid flow characteristics through tubes induced with multiple plain twisted tape inserts using water and 0.1 vol.% TiO_2/water nanofluid were investigated experimentally and numerically. The investigations were carried out for turbulent flow with Reynolds number ranging between 5000 and 20,000. The findings may be summarized as follows:

- It is found that tube fitted with quintuple twisted tape inserts performed highest in augmenting heat transfer. Nusselt numbers obtained by using quintuple twisted tapes inserts are up to 110.8% for nanofluid flow and 100.0% for water flow higher than in plain tube with water flow.

- As the number of twisted tapes increases, higher Nusselt number is obtained, compared to plain tube.

- The pressure drop and friction factor increase significantly by using twisted tape inserts and slightly by nanofluid.

- Using nanofluids with twisted tape inserts are resulting in more heat transfer augmentation than using each one individually.

- Using twisted tape inserts need to increase fluid pumping power due to high pressure drop produced. When using nanofluid without inserts, pumping power is not affected as the increase in the pressure drop is negligible compared to the case of pure water flow.

Accordingly, the combined technique of twisted tape inserts and nanofluid is recommended for heat exchanger application, as the heat transfer is considerably enhancing, with reasonable penalty in pumping power.

Acknowledgements

The authors acknowledge the supports provided by both, Al-Mustansiriyah University, Baghdad, Iraq for providing all the logistic and technical supports to produce this research using the relevant labs in the mechanical engineering department and Universiti Teknologi PETRONAS—Malaysia for providing the financial support to publish the research results using the budget under grant YUTP—FRG, CS: 015LC0-026.

Abbreviations

D	tube diameter (mm)
f	friction factor

I	turbulence intensity
K	thermal conductivity (W/m•K)
L	tube length (mm)
Nu	Nusselt number
Δp	pressure drop (Pa)
Pr	Prandtl number
$\bar{\bar{q}}$	heat flux (W/m^2)
Re	Reynolds number
V	mean velocity (m/s)
w	tape width (mm)
y	tape pitch 180° (mm)
δ	tape thickness (mm)
v	kinematic viscosity (m^2/s)
φ	nanoparticles volume fraction (%)
ρ	fluid density (kg/m^3)
μ	fluid dynamic viscosity (kg/m s)
c_p	fluid-specific heat (J/kg•K)

Subscripts

b	bulk
h	hydraulic
in	inlet
nf	nanofluid
s	surface
w	water

Acronyms

PT	plain tube
STT	tube with single twisted tape
TTT	tube with triple twisted tapes
QTT	tube with quintuple twisted tapes

Author details

Hussain H. Al-Kayiem[1*], Muna S. Kassim[2] and Saud T. Taher[2]

1 Mechanical Engineering Department, Universiti Teknologi PETRONAS, Perak, Malaysia

2 Mechanical Engineering Department, Al-Mustansiriyah University, Baghdad, Iraq

*Address all correspondence to: hussain_kayiem@utp.edu.my

References

[1] Kakaç S, Liu H. Heat Exchangers: Selection, Rating and Thermal Design. 2nd ed. USA, New York: CRC Press LLC, Taylor & Francis Inc.; 2002

[2] Kumar CN, Murugesan P. Review on twisted tapes heat transfer enhancement. International Journal of Science & Engineering Research. 2012; **3**(4):1-9

[3] Al-Kayiem HH, Saw CL, Afolabi L. Review on nanomaterials for thermal energy storage technologies. Journal of Nanoscience and Nanotechnology. 2013;**3**:60-71. DOI: 10.2174/ 22113525113119990011

[4] Dey D, Kumar P, Samantaray S. A review of nanofluid preparation, stability, and thermo-physical properties. Heat Transfer - Asian Research. 2017;**46**(8):1413-1442. DOI: 10.1002/htj.21282

[5] Ali N, Teixeira JA, Addali A. A review on nanofluids: Fabrication, stability, and thermophysical properties. Journal of Nanomaterials. 2018;**2018**: 6978130. DOI: 10.1155/2018/6978130

[6] Narayanan MV, Rakesh SG, Nanofluids SG. A review on current scenario and future prospective. IOP Conference Series: Materials Science and Engineering. 2018;**377**:012084. DOI: 10.1088/1757-899X/377/1/012084

[7] Said Z, Saidur R, Hepbasli A, Rahim NA. New thermophysical properties of water based TiO_2 nanofluid-the hysteresis phenomenon revisited. International Communications in Heat and Mass Transfer. 2014;**58**:85-95. DOI: 10.1016/j. icheatmasstransfer.2014.08.034

[8] He Y, Men Y, Zhao Y, Lu H, Ding Y. Numerical investigation into the convective heat transfer of TiO_2 nanofluids flowing through a straight tube under the laminar flow conditions. Applied Thermal Engineering. 2009;**29**: 1965-1972. DOI: 10.1016/j. applthermaleng.2008.09.020

[9] Bergles AE, Manglik RM. Heat transfer and pressure drop correlations for twisted tape inserts in isothermal tube: Part II-transition and turbulent flows. Transaction of ASME Journal of Heat Transfer. 1993;**115**:890-896. DOI: 10.1115/1.2911384

[10] Bhattacharyya S, Saha S, Saha SK. Laminar flow heat transfer enhancement in a circular tube having integral transverse rib roughness and fitted with centre-cleared twisted-tape. Experimental Thermal and Fluid Science. 2013;**44**:727-735. DOI: 10.1016/ j.expthermflusci.2012.09.016

[11] Bhattacharyya S, Chattopadhyay H, Benim AC. Simulation of heat transfer enhancement in tube flow with twisted tape insert. Progress in Computational Fluid Dynamics An International Journal. 2017;**17**(3):193-197. DOI: 10.1504/PCFD.2017.084356

[12] Bhattacharyya S, Chattopadhyay H, Benim AC. Computational investigation of heat transfer enhancement by alternating inclined ribs in tubular heat exchanger. Progress in Computational Fluid Dynamics, An International Journal. 2017;**17**(6):390-396. DOI: 10.1504/PCFD.2017.088818

[13] Bhattacharyya S, Benim AC, Chattopadhyay H, Banerjee A. Experimental investigation of heat transfer performance of corrugated tube with spring tape inserts. Experimental Heat Transfer. 2018;**32**(5):411-425. DOI: 10.1080/08916152.2018.1531955

[14] Pourrajabi M, Nezhad AR, Pourrajabi MR. Numerical study of turbulent flow and heat transfer of a nanofluid in a circular tube with twisted

tape insert. In: International Symposium on Advances in Science and Technology. Iran: 7th SASTech; 2013. DOI: 10.1007/s10973-017-6900-5

[15] Maddah H, Farokhi M, Aghayari R, Jahanizadeh S, Ashtary K. Effect of twisted-tape turbulator and nanofluid on heat transfer in a double pipe heat exchanger. Journal of Engineering. 2014;**2014**:920970. DOI: 10.1155/2014/920970

[16] Safikhani H, Eiamsa-ard S. Multi-objective optimization of TiO_2-water nanofluid flow in tubes fitted with multiple twisted tape inserts in different arrangement. Journal of Transport Phenomena in Nano-Micro Scale. 2015; **3**(2):89-99. DOI: 10.7508/tpnms.2015.02.003

[17] Saleh FA. Computational and experimental investigations on thermal and fluid flow characteristics for different models of tapes inserts with TiO_2/water nanofluid under turbulent conditions. International Journal of Mechanical & Mechatronics Engineering. 2018;**19**(2):1-14

[18] Eiamsa-ard S, Wongcharee K. Convective heat transfer enhancement using Ag-water nanofluid in a micro-fin tube combined with non-uniform twisted tape. International Journal of Mechanical Sciences. 2018;**146–147**: 337-335. DOI: 10.1016/j.ijmecsci.2018.07.040

[19] Xu C, Xu S, Wei S, Chen P. Experimental investigation of heat transfer for pulsating flow of GOPs-water nanofluid in a microchannel. International Communications in Heat and Mass Transfer. 2020;**110**:104403. DOI: 10.1016/j.icheatmasstransfer.2019.104403

[20] Chun BH, Kang HU, Kim SH. Effect of alumina nanoparticles in the fluid on heat transfer in double-pipe heat exchanger system. Korean Journal of Chemical Engineering. 2008;**25**(5): 966-971. DOI: 10.1007/s11814-008-0156-5

[21] Duangthongsuk W, Wongwises S. Effect of thermophysical properties models on the predicting of the convective heat transfer coefficient for low concentration nanofluid. International Communications in Heat and Mass Transfer. 2008;**35**(10): 1320-1326. DOI: 10.1016/j.icheatmasstransfer.2008.07.015

[22] Santra AK, Sen S, Chakraborty N. Study of heat transfer due to laminar flow of copper-water nanofluid through two isothermally heated parallel plates. International Journal of Thermal Sciences. 2009;**48**(2): 391-400. DOI: 10.1016/j.ijthermalsci.2008.10.004

[23] Namburu PK, Das DK, Tanguturi KM, Vajjha RS. Numerical study of turbulent flow and heat transfer characteristics of nanofluids considering variable properties. International Journal of Thermal Sciences. 2009;**48**(2):290-302. DOI: 10.1016/j.ijthermalsci.2008.01.001

[24] Available from: https://www.visualcapitalist.com/ [Retrieved on: 26 May 2020]

[25] Pak BC, Cho YI. Hydrodynamic and heat transfer study of dispersed fluids with submicron metallic oxide particles. Experimental Heat Transfer 1998; **11**(2):151-170. DOI:10.1080/08916159808946559

[26] Batchelor GK. The effect of Brownian motion on the bulk stress in the suspension of spherical particles. Journal of Fluid Mechanics. 1977;**83**(1):97-117. DOI: 10.1017/S0022112077001062

[27] Incropera FP, DeWitt PD, Bergman TL, Lavine AS. Fundamentals of Heat and Mass Transfer. USA, New York: John-Wiley & Sons Inc.; 2006

[28] Russo F, Basse NT. Scaling of turbulence intensity for low-speed flow in smooth pipes. Flow Measurement and Instrumentation. 2016;**52**:101-114. DOI: 10.1016/j.flowmeasinst.2016.09.012

Chapter 3

Tubular Heat Exchanger Fouling in Phosphoric Acid Concentration Process

Rania Jradi, Ali Fguiri, Christophe Marvillet and Mohamed Razak Jeday

Abstract

Fouling in phosphoric acid concentration is a persistent operational problem that compromises energy recovery in this process. Progress is hampered by the lack of quantitative knowledge of fouling dynamic effects on heat exchanger transfer. The object of this work is an experimental determination of the thermal fouling resistance in the tubular heat exchanger of phosphoric acid preheated installed in phosphoric acid concentration process. By measuring the inlet and outlet temperatures of phosphoric acid, steam temperature, suction and discharge pressure of the pump and acid density measurement, the overall heat transfer coefficient has been determined. The determination of the overall heat transfer coefficient with clean and fouled surfaces, allowed calculating the fouling resistance. The results from the heat exchanger studies showed that the fouling resistance increased with time and presented an asymptotic evolution in compliant with the proposed model by Kern and Seaton, with the existence of fluctuation. The poorly cleaned heat exchanger implied the absence of the induction period and caused, consequently, high values of the fouling resistance in a relatively short-time period.

Keywords: fouling, heat exchanger, heat transfer coefficient, fouling resistance, phosphoric acid concentration unit

1. Introduction

The main issues which various industries are facing are the accumulation of undesired substances or materials dissolved or presented as a suspension in the fluid on the heat transfer surfaces [1]. This phenomenon which is called as fouling affects the equipment operation by reducing their thermal effectiveness. This causes a significant economic loss due to the installations of regular cleaning [2, 3].

Fouling in heat transfer process is often inevitable and reduces energy efficiency and plant operability. Mitigation of fouling, and effective cleaning strategies, both require understanding the mechanisms involved in deposition and cleaning [4]. Many researches on fouling in heat transfer processes are dealt with, by reducing the efficiency of heat transfer and limiting productivity [5]. Phosphoric acid fouling in concentration process preheat exchangers is a persistent operational problem that compromises energy recovery in these process. Progress is hampered by the lack of quantitative knowledge of the dynamic effects of fouling on heat transfer exchanger [6]. Generally, phosphoric acid, which is the cold fluid, flows through the tube side while steam, which is the hot stream, flow through the shell side in heat exchangers [7]. The solution of concentrated phosphoric acid is supersaturated with calcium sulfate, resulting in the deposition on the contact material [8]. Given that the thermal conductivity of these scales is low, even a thin layer of scale can drastically reduce the overall heat transfer coefficient [9]. Furthermore, fluorosilicate and fluoroaluminate deposits on the acid ducts of clarifier tanks and evaporators can be imbedded in gypsum scale, which reduces pipe diameter and flow rate. In spite of considerable research efforts at the phosphoric acid type scale, no viable commercial solution has been found [10–12]. Behbahani et al. [13] have done a high number of fouling experiments in a side-stream of a phosphoric acid plant for various flow velocities, surface temperatures and concentrations in order to determine the mechanisms which control the deposition process. After identifying the effects of operational parameters on the deposition process, a fouling kinetic model by crystallization has been developed in Behbahani et al. [8]. A mathematical model has been elaborated to predict the fouling resistance in concentrating phosphoric acid [14]. The predicted fouling resistances were compared with the experimental data. Majority engineering calculations on heat transfer use the experimental heat transfer coefficients [15].

In this survey, we will examine the fouling phenomenon of the heat exchanger tubes for the preheat circuit of the phosphoric acid. The heat exchanger used for heating phosphoric acid is exposed to the fouling problem at the tube side of heat exchangers. In this context an experimental determination of the thermal fouling resistance by measuring the inlet and outlet temperatures of phosphoric acid, the temperature of steam, suction and discharge pressure of the pump and acid density measurement, the overall heat transfer coefficient has been determined. The determination of the overall heat transfer coefficient for the heat exchanger with clean and fouled surfaces makes it possible to calculate the fouling resistance.

2. Fouling mechanisms

Fouling can be divided into a number of distinct mechanisms. In general, many of these fouling mechanisms occur at the same time and each requires a different prevention technique. Among these different mechanisms, some represent different stages of the fouling process. The main mechanisms or stages of fouling include:

1. Period of initiation or delay. This is the clean surface period before dirt accumulation. This accumulation of relatively small deposits can even improve heat transfer over a clean surface and give the appearance of a "negative" fouling rate and a total negative fouling amount.

2. Particle fouling and formation, aggregation and flocculation.

3. Mass transport and migration of fouling agents to fouling sites.

4. Separation and deposition phase involving nucleation or initiation of fouling sites and attachment leading to deposit formation.

5. Growth, aging and hardening and increase of deposit resistance or auto-retardation, erosion and elimination.

3. Rate of fouling

Fouling is defined as a phenomenon that occurs with or without a temperature gradient in many natural, domestic and industrial processes. A surface is "dirty" when unwanted material accumulates there.

The fouling rate is normally defined as the average deposit surface loading per unit of surface area in a unit of time. Depending on the fouling mechanism and conditions, the fouling rate may be linear, falling, asymptotic or saw-tooth, as the case may be. **Figures 1** and **2** shows the different types of fouling rate.

1. Linear fouling is the type of fouling where the rate of fouling can be stable over time with the increase of fouling resistance and deposit thickness. It usually occurs when the temperature of the deposition in contact with the flowing fluid remains constant.

Ebert and Panchal [16] presented a fouling model expressing the average (linear) fouling rate under given conditions following two competing terms, namely a deposit term and an attenuated term.

$$\text{Fouling rate} = (\text{deposit term}) - (\text{anti} - \text{deposit term})$$

$$\frac{dRf}{dt} = \alpha \ Re^{\beta} \ Pr^{\delta} \exp\left(\frac{-E}{RT_{film}}\right) - \gamma\tau_w \quad (1)$$

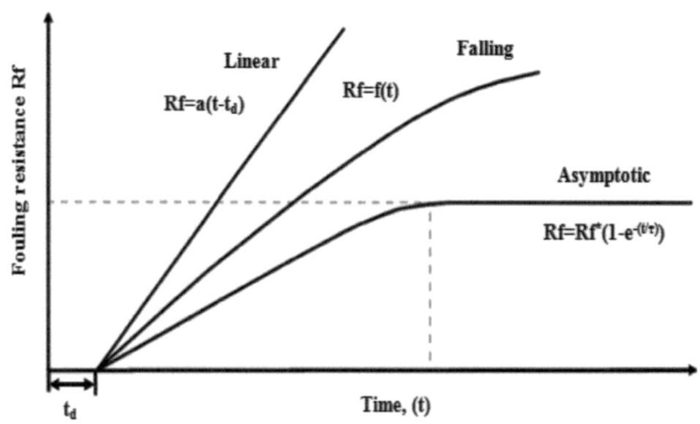

Figure 1.
Fouling curves.

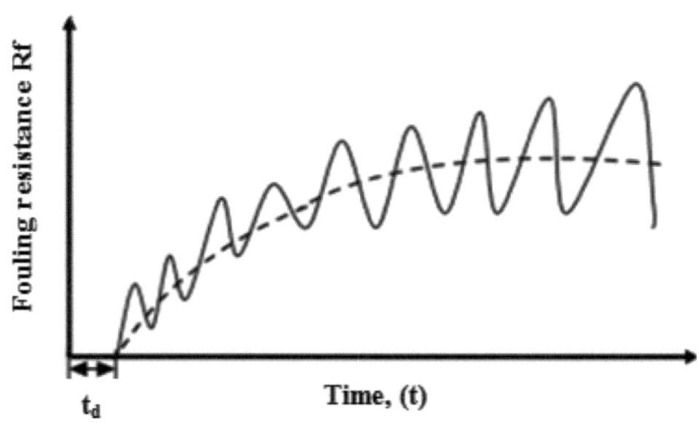

Figure 2.
Practical fouling curve.

where α, β, γ and δ are parameters determined by regression, τ_w is the shear stress at the tube wall and T_{film} is the temperature of the fluid film (average of local bulk and local wall fluid temperatures). The relationship in Eq. (1) indicates the possibility of identifying combinations of temperature and velocity below where fouling rates will be negligible. Ebert and Panchal [16] present this as the "threshold condition." The model in Eq. (1) suggests that the geometry of the heat exchanger which affects the surface and film temperatures, velocities and shear stresses can be effectively applied to maintain the conditions below "threshold conditions" in a given heat exchanger.

1. Falling fouling is the type of fouling where the fouling rate decreases with time, and the deposit thickness does not reach a constant value, although the fouling rate never drops below a certain minimum value. In general falling fouling is due to an increase of removal rate with time. Its progress can often be described by two numbers: the initial fouling rate and the fouling rate after a long period of time.

2. Asymptotic fouling rate is where rate decreases with time until it becomes negligible after a period of time when the deposition rate becomes equal to the deposit removal rate and the deposit thickness remains constant. This type of fouling generally occurs where the tube surface temperature remains constant while the temperature of the flowing fluid drops as a result of increased resistance of fouling material to heat transfer. Asymptotic fouling may also result from soft or poorly adherent suspended solid deposits upon heat transfer surfaces in areas of fast flow where they do not adhere strongly to the surface with the result that the thicker the deposit becomes, the more likely it is to wash off in patches and thus achieve some average asymptotic value over a period of time. The asymptotic fouling resistance increases with increasing particle concentration and decreasing fluid bulk temperature, flow velocity, and particle diameter. The asymptotic fouling model was first described by Kern and Seaton [17]. In this model, the competing fouling mechanisms result in asymptotic fouling resistance beyond which any additional increase in fouling does not happen.

3. Saw-tooth fouling occurs where part of the deposit is detached after a critical residence time or once a critical deposit thickness has been reached. The fouling layer then builds up and breaks off again. This periodic variation could be due to pressure pulses, scaling, trapping of air inside the surface deposits during shutdowns or other reasons. It often corresponds to the moments of system shutdowns, startups or other transients during operation.

4. Fouling measurement and monitoring

The fouling resistances can be measured experimentally or analytically. The main measurement methods include:

1. Direct weighing: the simplest method for assessing the extent of deposition on laboratory test surfaces is to weigh directly. The method requires an exact balance in order to be able to detect relatively small changes in the mass of deposits. It may be necessary to use thin walled tube to reduce the tare mass in order to increase the accuracy of the method.

2. Thickness measurement: in many examples of fouling the thickness of the deposit is relatively small, perhaps less than 50 μm, so that a direct measurement is not easy to obtain. A relatively simple technique provided there is reasonable access to the deposit, consists in measuring the thickness. By using a removable coupon or plate, the thickness of a hard deposit such as a scale, can be obtained using a micrometer or traveling microscope. For a deformable deposit containing a large proportion of water, e.g., a biofilm it is possible to use an electrical conductivity technique.

3. Heat transfer measurements: in this method, the fouling resistance can be determined according to the changes in heat transfer during the deposition process. The equation for the following operations will be Eq. (11). The data can be reported in terms of changes in overall heat transfer coefficient. A major hypothesis of this method is that the presence of the deposit does not affect the hydrodynamics of the flowing fluid. However, during the first stages of deposition, the surface of the deposit is generally rougher than the metal surface so that the turbulence in the fluid is greater than when it is flowing on a smooth surface. As a result the fouling resistance calculated from the data will be lower than if the increased turbulence level had been taken into account. It is possible that the increased turbulence offsets the thermal resistance of the deposit and negative values of thermal resistance will be calculated.

4. Pressure drop: as an alternative to direct heat transfer measurements it is possible to use changes in pressure drop caused by the presence of the deposit. The pressure drop is increased for a given flow rate due to the reduced flow area in the fouled condition and the roughness of the deposit. The shape of the curve relating pressure drop with time will generally, follow an asymptotic shape so that the time to achieve asymptotic fouling resistance can be determined. The method is often associated with the direct measurement of thickness of the deposit layer. Friction factor changes can also be used to indicate fouling of a flow channel.

5. Other techniques for assessing fouling: with regard to their effect on heat exchanger performance the measurement of heat transfer reduction or increase in pressure drop provide a direct indication. The simple methods of measuring deposit thickness described above are useful, but in general they require that the experiment be completed in order to allow access to the test sections. Ideally non-intrusive techniques would allow further deposition while maintaining experimental conditions without disturbance. Such techniques include the use of radioactive tracers and optical methods. Laser techniques can be used to study the accumulation and removal of deposits. In addition, infra-red systems are used to study the development and removal of biofilms from tubular test sections. Microscopic examination of deposits may provide further evidence of the mechanisms of fouling, but this is usually a “backup” system rather than providing quantitative data.

5. Prediction of fouling factor

As noted above, fouling has the effect of forming on the heat transfer surface a substantially solid deposit of low thermal conductivity, through which heat is to be transferred by conduction. But as the thermal conductivity of the fouling layer and its thickness are not generally known, the only possible solution to the heat transfer problem is to introduce a fouling factor to take into account the additional resistance to heat transfer and possible calculation of the overall coefficient of heat transfer. A fouling coefficient is also sometimes specified, it is the reciprocal value of the fouling factor. When carrying out heat transfer calculations, the selection of fouling factors must be made with caution, especially when the fouling resistances completely dominate the thermal design.

The influence of inherent uncertainties in fouling factors is generally greater than that of uncertainties in other design parameters such as fluid properties, flow rates and temperatures [18]. An important fouling factor is sometimes adopted as a safety margin to cover uncertainties on the properties of fluids and even in the knowledge of the process, but the use of an excessively large fouling factor will result in an oversized heat exchanger with two or three times more area than is necessary. Although many tabulations based on the experiment are available and provide typical fouling factors such as the TEMA RGP-T-2.4 table [19], an acceptable assessment of the effects of fouling needs to be judged and evaluated for each particular application. Such tabulations can, however, serve as a guide in the absence of more specific information.

A number of semi empirical models have been proposed over the years for the prediction of the rate of fouling in heat exchangers or for estimating a fouling factor to be used in heat transfer calculations.

The first work on this subject began in the late 1950s with Kern and Seaton [17].

The modeling resulting from this work is based on the assumption that two processes act simultaneously. The first process is that of particle deposition characterized by a deposition flux that is constant if the concentration is also constant. The second process is that of the re-entrainment of particles characterized by a re-entrain flow ϕ_r dependent on the mass of particles (m_p) deposited. The particle balance of the deposit is expressed according to the following equation:

$$\frac{dm_p}{dt} = \phi_d - \phi_r \tag{2}$$

The deposition process is designed as the serialization of particle transport and adhesion mechanisms. The following assumptions are made:

- consideration of a single type of fouling;
- homogeneity of the deposit;
- not taken into account of the phase of initiation of the deposit and the state of surface;
- constancy of the properties and thermo-physical characteristics of the fluid and the deposit.

The particle wall transport phase controls the deposition process while the shear stress controls the re-entrain phase of the particles. Thus, considering the proportionality of ϕ_d as a function of the deposited mass of particles, we can write the following equations:

$$\phi_d = k_p * (C_b - C_w) \tag{3}$$

$$\phi_r = C_1 * \tau_w * m_p \tag{4}$$

Or
k_p is the transport coefficient.
C_b the particle concentration in the fluid.
C_w the particle concentration at the wall.
C_1 a dimensional constant.
τ_w the shear stress exerted by the fluid on the deposit.
Equation (2) thus becomes:

$$\frac{dm_p}{dt} = k_p * (C_b - C_w) - C_1 * \tau_w * m_p \tag{5}$$

The solution of Eq. (5) is thus:

$$m_p = \frac{K_p * (C_b - C_w)}{C_1 * \tau_w} * [1 - \exp(-C_1 * \tau_w * t)] \tag{6}$$

Assuming that $\tau = \frac{1}{C_1 * \tau_w}$ and $m_p^* = \frac{K_p * (C_b - C_w)}{C_1 * \tau_{w.}}$
We can thus express the equation as follows:

$$m_p = m_p^* * \left[1 - exp\left(-\frac{t}{\tau}\right)\right] \tag{7}$$

Considering that the initial fouling flow is equal to the deposition flow and that the thermo physical properties of the deposit (conductivity and density) are constant, it is thus possible to express Eq. (7) in the form of a thermal fouling resistance:

$$Rf(t) = Rf^* * \left[1 - exp\left(-\frac{t}{\tau}\right)\right] \tag{8}$$

With $Rf(t)$, the evolution of the fouling resistance as a function of time expressed in [m^2-K/W] Rf^*, the asymptotic value of the fouling resistance expressed in [m^2-K/W] (this value characterizes the situation where the deposition rate equals the breakout speed). t, the time expressed in [s] τ, the characteristic time expressed in [s] and whose value is generally attributed to the time required for the fouling resistance to reach its asymptotic value if the evolution of this kinetics was linear.

The Kern and Seaton [17] model therefore provides a mathematical description of the concept of simple fouling. This equation verifies the asymptotic behavior of the formation of a particulate deposit on the exchange surface of a heat exchanger. All models and theory of fouling are based on this model.

An apparent weakness of the Kern and Seaton [17] model is that the re-entrain flow depends on the thickness of the deposition layer. As a result, it is only once a significant deposit thickness has accumulated that the role of the re-entrain term becomes significant [20]. In the case of high speed flow, the deposit would be completely removed.

We also note that this model requires to go back on the values of Rf^* and τ. In general, there is no way to predict these values unless detailed experimental work has been done [21]. These values are thus established on a given installation and especially under given operating conditions. Thus, a modification, even minor, of the operating conditions (for example, water quality, flow modification) significantly modifies the parameters of the model and leads to a bad modeling of the fouling [22].

However, we note various works that make it possible to know the impact of the flow velocity (Um) on the value of the asymptotic fouling resistance Rf^*, or even to determine the value Rf^*.

Different authors thus propose a relationship of proportionality of type:

$$Rf^* \sim Um_i \tag{9}$$

With regard to the tube exchangers: for Kern and Seaton [17] the value of i is −1. For Watkinson [23], this constant takes the value of −1.2 to −2.

As far as plate heat exchangers are concerned, Muller-Steinhagen [24] has in its study demonstrated a relation of proportionality between the asymptotic resistance of fouling Rf^* and the inverse of the speed squared (i.e., an exponent $i - 2$ in Eq. (9)), without providing a general relationship.

In the same context, Grandgeorge [25] proposes an empirical relation resulting from several experiments on different industrial size plate heat exchangers. In this context, Grandgeorge [25] established that the use of the initial pressure drop in the heat exchanger (ΔPo) in place of the flow velocity makes it possible to correlate with the aid of a single relationship the value of the asymptotic resistance Rf^*. The relationship is as follows:

$$Rf^* = \frac{1}{4 * \Delta Po} \tag{10}$$

Based on these observations, this model has been revised and modified by various researchers with various descriptions of the term deposition and re-entrain: Only empirical parameters were added and derived solely from the experimental study [20, 25].

6. Experimental procedure

The phosphoric acid concentration loop is allowed to concentrate—by evaporation—the phosphoric acid from 28 to 54% P_2O_5 in a forced-circulation evaporator closed loop, functioning under vacuum feeded by a barometric condenser. The system used for concentration composed of a stainless steel tubular heat exchanger, a centrifugal pump, a boiler or expansion chamber, a barometric condenser and a basket filter [26].

The inclusion of the dilute acid is done at the basket filter where it mixes with the circulating acid in order to protect the pump from abrasion and to limit the heat exchanger fouling, which reduces the stop frequency for washing. The circulation pump then aspirates the mixture formed and sends it to the inlet of the heat exchanger at a temperature in the order of 70°C. The heat exchanger allows heating the phosphoric acid at a temperature in the order of 80°C. The steam undergoes a condensation at a temperature of around 120°C at the level of the exchanger. The condensate will be sent to a storage tank before being returned to the utility center.

The overheated mixture of the acid outgoing the exchanger then passes into the boiler where an amount of water evaporates and the production of concentrated acid is done by overflow in inner tube of the boiler and the rest will be recycled. The condenser also ensures the re-entrain of incondensable outgoing of the boiler by the effect of water tube created by the waterfall. At the foot of the barometric guard, the seawater is gathered in a guard tank before being rejected to the sea (**Figure 3**).

Our experimental study is based on the following hypotheses.

1. The flow of two fluids (Phosphoric acid and steam) is at counter current.

2. Values of the thermo-physical properties of the fluids were considered constant.

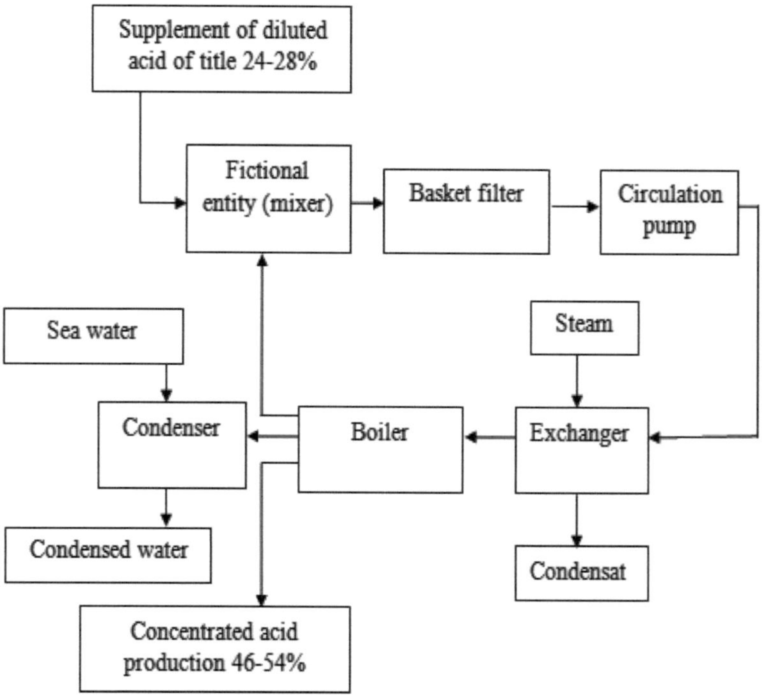

Figure 3.
Simplified diagram of the phosphoric acid concentration unit.

3. The thermal losses were neglected.

4. The inlet and outlet temperatures of the two fluids are determined at the extremities of the heat exchanger.

5. Pump suction and discharge pressure measurements are taken at the extremities of the circulation pump.

7. Calculation method

The experimental data was collected out during 1 year. The method that we used to follow the fouling evolution consists in carrying out a heat balance at the boundaries of the heat exchanger by the intermediary of measurements of the inlet and outlet temperatures pump suction and discharge pressure measurements and acid density measurement (**Figure 4**). The latter was taken each 2 h during all the day.

This method, albeit indirect, makes it possible to detect the necessary moment to shut down the installation for cleaning. In the current study, the temporal evolution of the fouling resistance of the phosphoric acid was studied.

The calculation of the fouling resistance was done using the following relation:

$$Rf(t) = \frac{1}{U_s} - \frac{1}{U_p} = \frac{1}{U(t)} - \frac{1}{U(t=0)} \quad (11)$$

7.1 Calculation of U_s

The overall heat transfer coefficient at the dirty state was given in the time course, via the expression:

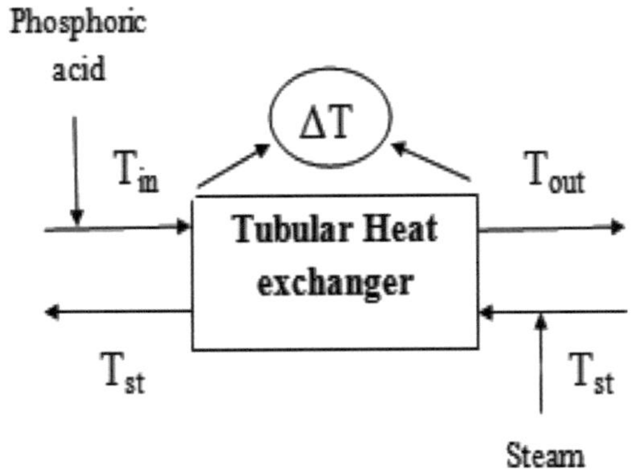

Figure 4.
The measurement method at the boundaries of the heat exchanger.

$$U_s = U(t) = \frac{\dot{m}_{ac,cir} * Cp_{ac} * (T_{out,ac} - T_{in,ac})}{A * F * \Delta T_{ml}} \tag{12}$$

This relation is taken by the evaluation of energy on the heat exchanger by supposing the isolated system and the physical properties of the two fluids, as well as, the heat transfer coefficients stay constant along the exchanger.

7.2 Calculation of U_p

In the phosphoric acid concentration unit, the operating conditions at the limits of the heat exchanger unstable, it is necessary to disclose the heat exchange coefficients in the proper conditions U_p corresponding to the new operating conditions. Assuming that the cleaning between operational runs is perfect and that the heat exchangers are totally free of fouling at the beginning of a new run. The initial value of the overall heat transfer coefficient at the beginning of every cycle is considered as the value of the overall heat transfer coefficient in the clean state.

$$U_p = U(t = 0) = \frac{\dot{m}_{ac,cir} * Cp_{ac} * (T_{out,ac} - T_{in,ac})}{A * F * \Delta T_{ml}} \tag{13}$$

8. Results and discussion

8.1 Temporal evolution of the fouling resistance

The evolution of the fouling resistance in the phosphoric acid concentration process heat exchanger was followed for a study period quoted previously. This heat exchanger is already in service for 100 days before the be-ginning of the present study. However, it has carried out a stop that lasted 12 hours then its return to service. This stop is for the heat exchangers cleaning.

All the results from the fouling resistance are presented on **Figure 5**.

According to the values of these resistances, which are the majority higher than zero, this experimental data presents a fouling state. This is evident since, as mentioned before, this exchanger is in service for more than 3 months. The curves presented show that the temporal evolution of the fouling resistance, seems to follow an asymptotic evolution, which conforms to the model of Kern and Seaton [17], with the absence of the induction period. That can be explained to the rapid evolution of this phenomenon associated in particular with the characteristics of the

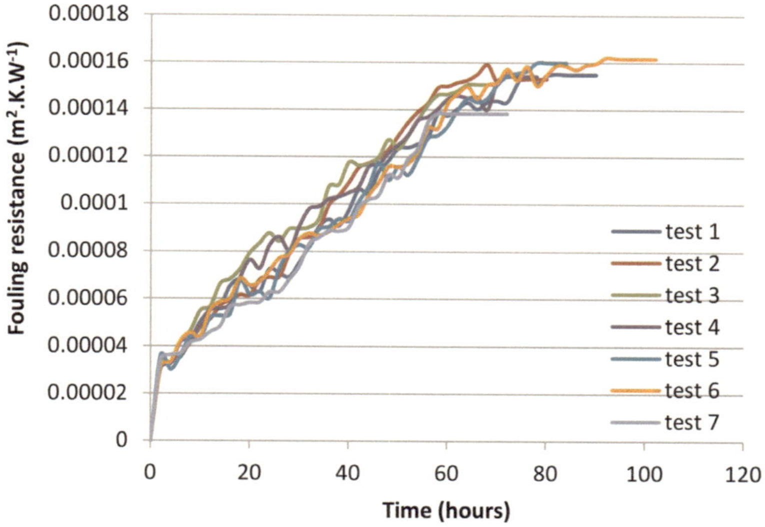

Figure 5.
Variation of the fouling resistance as a function of time for the stainless-steel-tubular heat exchanger.

treated phosphoric acid. As it appears clearly as the fouling resistance increases with the time until reaching a maximum value varied from $1.38 * 10^{-4}$ to $1.61 * 10^{-4}$ $m^2.K.W^{-1}$.

The series functioned for more than 5 days, a sufficient period so that the resistance asymptotic value is reached. The fluctuation observed on these curves are due to the variation of flow, which, acting on the shear stress to the wall, causes re-entrain deposit particles or their deposition according to the sent flow value.

8.2 Temporal evolution of the overall heat transfer coefficient

From Eq. (11), we notice that the overall heat transfer coefficient is inversely proportional to the fouling resistance.

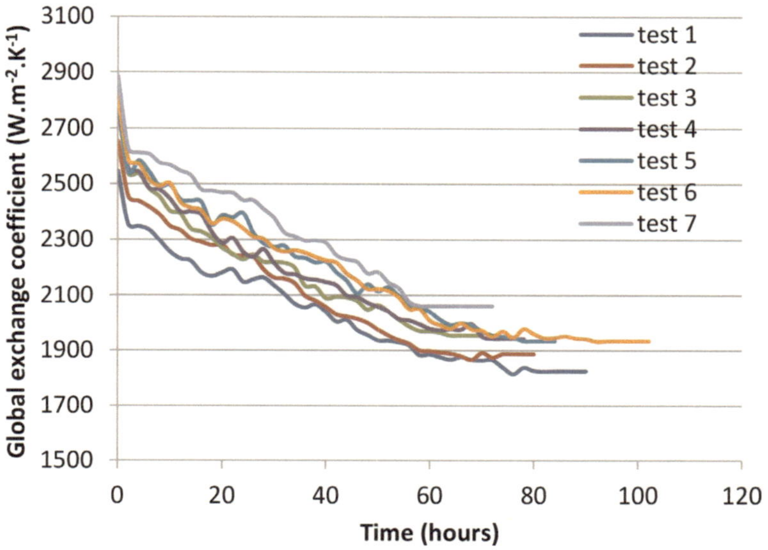

Figure 6.
Variation of the overall heat transfer coefficient as a function of time for the stainless-steel-tubular heat exchanger.

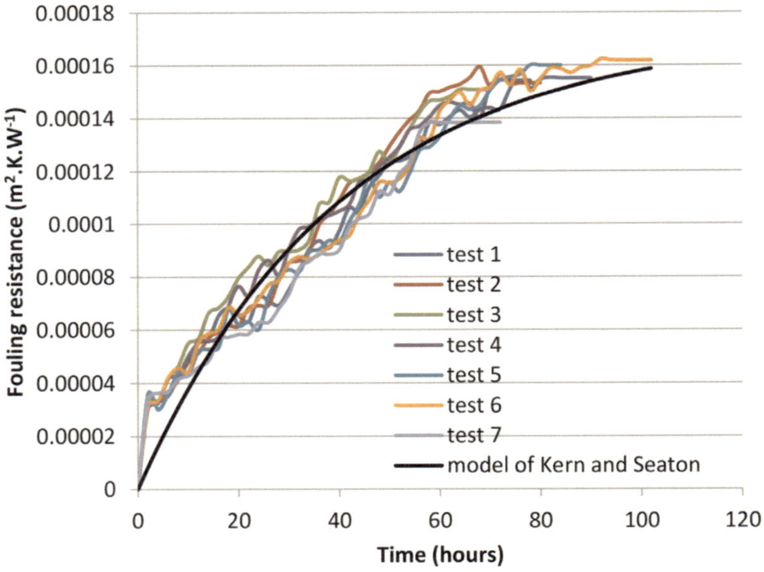

Figure 7.
Kinetics of fouling of the stainless-steel-tubular heat exchanger.

The fouling resistance increases over time, which leads to a decrease in the flow of heat exchanged between the phosphoric acid and the steam, and subsequently the decrease in the overall heat transfer coefficient. As it appears clearly on **Figures 5** and **6**, when the fouling resistance increases with the time, the overall heat transfer coefficient decreases until reaching a minimum value varied from 1821 to 2078 $W.m^{-2}.K^{-1}$.

8.3 Temporal evolution of the fouling resistance obtained from both measurement and the Kern and Seaton model

One of the earliest correlative models for the characterization of the asymptotic kinetics of fouling, we distinguish Kern and Seaton [17], whose general expression is as follows:

$$Rf(t) = Rf^* * \left[1 - exp\left(-\frac{t}{\tau}\right)\right] \tag{14}$$

This model gives rather satisfactory results, provided that the asymptotic value of the thermal resistance Rf^* as well as the time constant τ are evaluated, which strongly conditions the accuracy of the model.

The analysis of the experimental data which makes it possible to carry out the plots of **Figure 7** gives us the results of the two greatness Rf^* and τ for the stainless steel tubular heat exchanger. The asymptotic model is fairly faithful to the experimental data with determination coefficient R^2 close to 1 (**Table 1**).

Rf^* [$m^2.K.W^{-1}$]	**τ [h]**	**R^2**
$1.72*10^{-4}$	40.32	0.975

Table 1.
Values of the asymptotic fouling resistance, the time constant and the determination coefficient for the stainless-steel tubular heat exchanger.

9. Conclusion

The monitoring of heat exchangers provides the sound knowledge of the fouling evolution in the specific conditions of the process. Deposit formation is a thermal resistance which leads important economic penalties.

The aim of this work was the study of the heat exchanger fouling phenomenon in the concentration phosphoric acid process. Secondly, the study concerned the temporal evolution of the fouling resistance and the overall heat transfer coefficient.

The results achieved revealed an asymptotic evolution of the fouling resistance, compliant with the model of Kern and Seaton with the lack of the induction period, which is explained by the consequence of an improper cleaning, or a deviation between the present study and the beginning of the heat exchangers functioning after the last stop.

Nomenclature

A	area, m^2
Cp	specific heat capacity, $J.Kg^{-1}.K^{-1}$
F	correction factor (=1 for a steam condenser)
$\dot{m}$	mass flow rate, $kg.s^{-1}$
P	pressure, bar
Q	thermal power, W
Rf	fouling resistance, $m^2.K.W^{-1}$
T	temperature, K
t	time, h
U	overall heat transfer coefficient, $W.m^{-2}.K^{-1}$
Greek letters	
Δ	difference of greatness between two points
τ	time required to reach 63.2% of Rf^*, h
Subscripts	
ac	acid
cir	circulation
ex	exchange
in	input
ml	logarithmic mean
0	clean
out	output
p	clean state
s	dirty state
st	steam
*	asymptotic value

Author details

Rania Jradi[1]*, Ali Fguiri[1], Christophe Marvillet[2] and Mohamed Razak Jeday[1]

1 Research Laboratory "Process, Energy, Environment and Electrical Systems", National Engineering School of Gabes (ENIG), Gabes, Tunisia

2 CMGPCE Laboratory, French Institute of Refrigeration (IFFI), National Conservatory of Arts and Crafts (CNAM), Paris, France

*Address all correspondence to: raniajradi@yahoo.fr

References

[1] Demasles H, Mercier P, Tochon P, Thonon B. Guide de L'encrassement des Echangeurs de Chaleur. France: Editions-GRETh; 2007

[2] Farajollahi B, Gh Etemad S, Hojjat M. Heat transfer of nanofluids in a shell and tube heat exchanger. International Journal of Heat and Mass Transfer. 2010;**53**:12-17

[3] Lalot S, Pálsson H. Detection of fouling in a cross-flow heat exchanger using a neural network based technique. International Journal of Thermal Sciences. 2010;**49**: 675-679

[4] Pogiatzis T, Ishiyama EM, Paterson WR, Vassiliadis VS, Wilson DI. Identifying optimal cleaning cycles for heat exchangers subject to fouling and ageing. Applied Energy. 2012;**89**: 60-66

[5] Mohanty DK, Singru PM. Use of C-factor for monitoring of fouling in a shell and tube heat exchanger. Energy. 2011;**36**:2899-2904

[6] Yeap BL, Wilson DI, Polley GT, Pugh SJ. Mitigation of crude oil refinery heat exchanger fouling through retrofits based on thermo-hydraulic fouling models. Chemical Engineering Research and Design. 2004;**82**:53-71

[7] Radhakrishnan VR, Ramasamy M, Zabiri H, Thanh V Do, Tahir NM, Mukhtar H, et al. Heat exchanger fouling model and preventive maintenance scheduling tool. Applied Thermal Engineering. 2007;**27**: 2791-2802

[8] Behbahani RM, Muller-Steinhagen H, Jamialahmadi M. Investigation of scale formation in heat exchangers of phosphoric acid evaporator plants. The Canadian Journal of Chemical Engineering. 2008;**84**:189-197

[9] Jamialahmadi M, Müller-Steinhagen H. Heat exchanger fouling and cleaning in the dihydrate process for the production of phosphoric acid. Chemical Engineering Research and Design. 2007;**85**:245-255

[10] Amjad Z. Calcium sulfate dihydrate (gypsum) scale formation on heat exchanger surfaces: The influence of scale inhibitors. Journal of Colloid and Interfacial Science. 1988;**123**: 523-536

[11] Geddert T, Bialuch I, Augustin W, Scholl S. Extending the induction period of crystallization fouling through surface coating. Heat Transfer Engineering. 2009;**30**:868-875

[12] Shih WY, Albrecht K, Glater J, Cohen Y. A dual-probe approach for evaluation of gypsum crystallization in response to antiscalant treatment. Desalination. 2004;**169**:213-221

[13] Behbahani RM, Muller-Steinhagen H, Jamialahmadi M. Heat exchanger fouling in phosphoric acid evaporators-evaluation of field data-. In: Proceedings EIC on Heat Exchanger Fouling and Cleaning: Fundamentals and Applications. Mexique: Santa Fé; 2003

[14] Behbahani RM. Heat Transfer and Heat Transfer Fouling in Phosphoric Acid Evaporators. European Union: Lap Lambert Academic Publishing; 2010

[15] Taler D. Experimental determination of correlations for average heat transfer coefficients in heat exchangers on both fluid sides. Heat and Mass Transfer. 2013;**49**: 1125-1139

[16] Ebert WA, Panchal CB. Analysis of Exxon crude-oil slip stream coking data. In: Fouling Mitigation of Industrial Heat-Exchange Equipment. New York: Begell House; 1996:451-460

[17] Kern D, Seaton R. A theoretical analysis of thermal surface fouling. British Chemical Engineering. 1959;**4**: 258-262

[18] Riverol C, Napolitano V. Estimation of the overall heat transfer coefficient in a tubular heat exchanger under fouling using neural networks. Application in a flash pasteurizer. International Communications in Heat and Mass Transfer. 2002;**29**:453-457

[19] Standards of Tubular Exchanger Manufacturers Association. Tubular Exchanger Manufacturers Association edition. 9th ed. New York, Tarrytown; 2007

[20] Wang Y, Yuan Z, Liang Y, Xie Y, Chen X, Li X. A review of experimental measurement and prediction models of crude oil fouling rate in crude refinery preheat trains. Asia-Pacific Journal of Chemical Engineering. 2015;**10**:607-625

[21] Bott TR. Fouling of Heat Exchangers. Science B.V edition. Amsterdam: Elsevier; 1995

[22] Weber C, Tremeac B, Marvillet C, Castelain C. Analysis of different fouling predictive models in a heat exchanger from experimental data. In: Proceedings of ECOS 2016—The 29th International Conference on Efficiency, Cost, Optimization, Simulation and Environmental Impact of Energy Systems. Slovenia: Portoroz; 2016

[23] Watkinson AP. Particulate fouling of sensible heat exchangers. Vancouver: The University of British Columbia; 1968

[24] Muller-Steinhagen H. Particulate fouling for low finned tubes and for plate heat exchangers. In: International Conference on Fouling in Process Plant. England: St Catherine College of Oxford; 1988

[25] Grandgeorge S. Contribution à l'étude de l'encrassement particulaire des échangeurs à plaques corruguées en milieu liquide. L'Institut National Polytechnique de Grenoble; 1996

[26] Becker P. Phosphates and Phosphoric Acid: Raw Materials, Technology and Economics of the Wet Process. 2nd ed. New York: CRC Press; 1989

Chapter 4

Fouling in Heat Exchangers

Sergio García and Alfredo Trueba

Abstract

A major problem in industries that use heat exchanger equipment cooled with water in their industrial processes is biofouling. In the design and operation of heat exchangers cooled with water, a coefficient of biological must be considered, which affects the efficiency of the equipment. For this reason, it is necessary to apply appropriate antifouling treatments to the design of each heat exchanger. In order to minimize the undesirable phenomenon of biofouling, various mitigation methods have been developed over the last 30 years, both online and offline, of a physical, chemical, or biological nature. Most of these methods are well contrasted and are applied in the regular operation of the facilities, although some methodology approaches are in the research and development phase. However, the application of most of these methods requires interrupting the production, periodically, in order to clean the biofouling, seriously damaging the performance and operation of the installation. The "online" methods to biofouling control are chemical (oxidizing, and non-oxidants), biological and physical treatments. Nowadays, other methodologies of biofouling mitigation that do not affect the environment are being investigated, although, until now, none have been found that are substitutes for chemical agents and that have the same or with higher efficiency.

Keywords: biofouling, biofilm, fouling, antifouling, energy, efficiency

1. Introduction

A major problem in industries that use heat exchange equipment cooled with seawater in their industrial processes is biological fouling (biofouling) (**Figure 1**). Biofouling acquires greater relevance when it develops in the marine environment, due to the high biological activity of seawater [1]. In the design and operation of heat exchangers cooled with seawater, a biological fouling coefficient must be considered, which affects the efficiency of the equipment [2]. For this reason, it is necessary to apply appropriate antifouling treatments to the design of each heat exchanger [3].

In order to minimize the undesirable phenomenon of biofouling, various mitigation methods have been developed over the last 30 years, both online and offline, of a physical, chemical, or biological nature. Most of these methods are well contrasted and are applied in the regular operation of the facilities, although some methodology approaches are in the research and development phase. However, the application of most of these methods requires interrupting production, periodically, in order to proceed with biofouling cleaning, as this seriously impairs the performance and operation of the facility (**Figure 2**).

Oxidizing chemical methods are available [4] and non-oxidants [5], biological [6–8] for "on-line biofouling control. Nowadays, other methodologies of biofouling mitigation that do not affect the marine environment are being investigated [3],

Figure 1.
Presence of mussels in the water box of a condenser (Source: By courtesy of Nuclenor).

Figure 2.
Tubular heat exchanger in maintenance due to the presence of biofouling in its tubular bundle (Source: By courtesy of Merus-the water conditioner company).

although, until now, none have been found that are substitutes for chemical agents and that have the same or with higher efficiency.

Other "offline" methods, such as antifouling coatings (AF) or mechanical cleaning, are widely used in heat exchangers. AF coatings have the drawback of using chemicals that can harm the environment. Nowadays, new coatings are being researched to avoid the growth of biofouling, without harming the environment [9, 10].

2. Classification of the fouling

For a better understanding of the biofouling problem, this chapter begin of the description general term "fouling" and its types, among which is biofouling. This term is used to define the accumulation of biotic and/or abiotic deposits on a surface in contact with a fluid. This type of soiling can occur in nature in many different ways and is usually classified in the following categories [11]:

- Biological fouling
- Fouling by chemical reaction
- Fouling by corrosion

- Ice-cold fouling
- Fouling by precipitation
- Fouling by particles

2.1 Biological fouling

The biological fouling or biofouling in the Anglo-Saxon literature depends on the material on which the accumulation of macrofouling or macrobial biofouling occurs, which originates from microfouling or microbial biofouling.

2.2 Fouling by chemical reaction

Deposits caused by chemical reaction can be initiated by the contact of two chemicals in a heat transfer process. In some cases, the metal surface favors precipitation by acting as a catalyst.

This type of fouling is frequently observed in the petroleum industry, in which the refining processes induce the polymerization of unsaturated hydrocarbons (**Figure 3**).

2.3 Fouling by corrosion

This type of fouling occurs as a result of electrochemical or microbial corrosion of a surface in contact with a fluid (**Figure 4**). The composition of corrosion products is normally a function of the material from which they originate. According to [12] corrosion must be taken into account in the design of heat exchangers in order to generate more information on the thermal resistance associated with fouling by corrosion.

2.4 Freezing fouling

It is the accumulation of solid deposits on a refrigerated surface due to the solidification of a fluid or one of its constituents. The accumulation of this type of fouling depends on the temperature of the surface and the shearing effect produced by the fluid on the surface as a function of its velocity of circulation [13, 14]. This phenomenon is seen in the solidification of the wax, whose melting point is high, in a hydrocarbon solution on a heat exchange surface (**Figure 5**).

Figure 3.
Residue of hydrocarbon in the tubes of a heat exchanger (Source: By courtesy of H&C Heat Transfer Solutions Inc. 2013).

Figure 4.
Oxides of corrosion in the tubes of a heat exchanger (H&C Heat Transfer Solutions Inc. 2013).

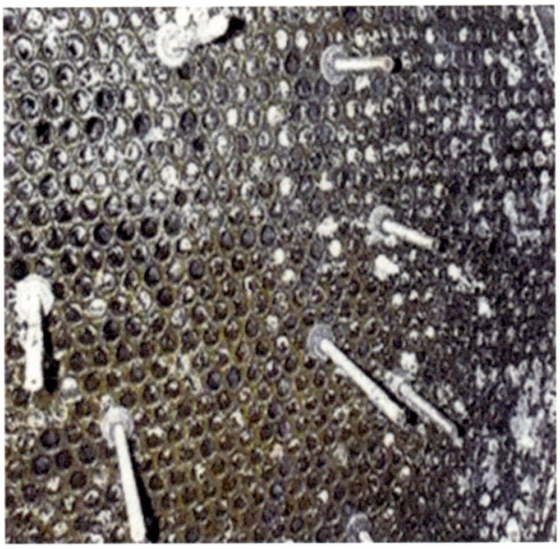

Figure 5.
Icy icing of polymer products in a heat exchanger (Source: By courtesy of H&C Heat Transfer Solutions In.).

2.5 Fouling by precipitation

The origin of this type of fouling is the precipitation or scaling of dissolved substances on the surfaces. Precipitation usually occurs from aqueous solutions based on their temperature and the solubility of dissolved substances, such as calcium carbonate ($CaCO_3$), whose precipitation of the solution occurs on a superheated surface when the water contained in the water evaporates (**Figure 6**).

The precipitation of the salts dissolved in the water originates when coming into contact with the wall of the hot tube and supersaturation occurs. The deposits adhered to the surface may have different hardness depending on their composition. Thus, when the deposit is only composed of salts, it is hard and very adherent. On the other hand, if the deposit is a mixture of salts and suspended solids, its hardness is low, and they adhere weakly [15].

The precipitates usually observed in industrial facilities that use aqueous solutions are the following [16]:

- Calcium carbonate
- Calcium sulfate
- Calcium oxalate
- Barium sulfate
- Magnesium hydroxide
- Silicates
- Aluminum oxides
- Aluminum silicate
- Copper
- Phosphates
- Magnetite

These precipitates are very common in boilers and heat exchangers that operate with water of high hardness.

2.6 Fouling by particles

It is the accumulation of solid particles that are suspended in a fluid (liquid or gas). The particles are deposited on surfaces by means of various mechanisms, some of which are closely linked to properties of particles as specific as coagulation. The amount of deposit and the rate of accumulation on the surfaces depend on the size and nature of the particles, as well as the environmental conditions. When gravity is the main mechanism of deposition of suspended particles on a surface, fouling is called sedimentation (**Figure 7**) [17].

Figure 6.
Precipitated salts in the tubes of a heat exchanger (Source: By courtesy of H&C Heat Transfer Solutions Inc).

Figure 7.
Accumulation of solid particles in the plate of a tubular heat exchanger (Source: By courtesy of AH&C Heat Transfer Solutions Inc.).

3. Biofouling

The term biofouling is clearly defined and accepted by the scientific community, although in the various definitions proposed by different authors some aspects of others are highlighted, depending on the importance given by each author to the different facets of the phenomenon. Below are several definitions of this term, published over the past years by different authors:

- Biofouling is simply the union of an organism or organisms to a surface in contact with water over a period of time [18].
- Biofouling is the colonization of submerged surfaces, produced by unwanted organisms such as bacteria, algae, and barnacles. Biofouling has detrimental effects on transport and leisure vessels, heat exchangers, oceanographic sensors, and aquaculture systems [19].
- Biofouling can be defined as the union and growth of a community of plants and animals [20].
- Biofouling is defined as the fixation and growth of organisms on submerged surfaces of an artificial nature. The accumulation of biofouling includes barnacles and other sessile marine invertebrates [21].
- Biofouling is the unwanted accumulation of microorganisms, algae, arthropods, or mollusks on a surface [18].

The definition of the term biofouling accepted is the following: "biofouling can be defined as the undesirable phenomenon of adhesion and accumulation of biotic deposits on an artificial surface submerged or in contact with seawater." This accumulation or embedding consists of an organic film composed of microorganisms embedded in a polymer matrix created by themselves (biofilm), where they can reach and retain inorganic particles (salts and/or corrosion products) as a result of other types of fouling developed in the process. This biofilm composed of microorganisms (microbial biofouling or microfouling) can lead to the accumulation of macroorganisms (macrobial biofouling or macrofouling) [22].

3.1 Evolution of the research of biofouling

Zobell [23] examined a natural marine population through a direct microscope and observed that the number of bacteria adhered to the surfaces was much higher than that found in the medium. In addition, these bacteria were characterized by showing an activity and high growth rates, concluding that the bacteria were attracted to the surfaces to which they adhered to form sessile populations.

The first detailed examination of the nature and composition of the biofilm had to wait for the appearance of the electron microscope, which provided a greater extension with respect to optical microscopy and, in its scanning and transmission modalities, was able to show the variety of microorganisms that made up the biofilm in a treatment plant [24]. Staining the biofilm with ruthenium red and fixing it with osmium tetroxide allowed to show that the material that surrounds and encloses the cells that compose it is composed mainly of polysaccharides. Costerton et al. [25], taking as a starting point the observations of sessile communities in the mountain streams, elaborated a theory that explains the mechanisms through which microorganisms adhere to living or inert materials, as well as the benefits obtained for their ecological niche. From this moment, numerous studies of the biofilm were developed in both industrial and ecological scenarios [26].

The possible effects of the biofilm in industrial processes are derived from the beneficial or harmful reactions that can be carried out by the microorganisms that compose it and that depend on the environmental conditions of the environment. These conditions have a great influence on the growth and metabolic activity of the biofilm [27].

The bacteria that make up the biofouling can be up to a thousand times more resistant to antibiotics than the same bacteria grown in a controlled liquid medium. The mechanisms responsible for this resistance include (1) the physical and chemical diffusion barrier that constitutes the matrix of the biofilm to the penetration of antimicrobials, (2) the slowed growth of biofilm bacteria due to nutrient limitation, (3) the existence of microenvironments that antagonize the action of the antibiotic, and (4) the activation of stress responses, which cause changes in the physiology of the bacteria and the appearance of a specific biofilm phenotype that actively combats the negative effects of antimicrobial substances [28]. Due to this resistance, antifouling substances must be highly effective and incorporated at considerably high concentrations, which can lead to harmful effects on the environment.

3.2 Microfouling

The biofilm (microfouling) is composed mainly of water, in intervals that are between 87 and 99% of the total content. The rest of the components is very variable depending on the medium in which it is studied [17]. The matrix of the biofilm is a complex formed mainly by exopolysaccharides and bacterial cells. Other compounds that can be found are macromolecules such as proteins, DNA, and various products derived from the destruction of the cell wall of bacteria [29].

3.2.1 Development of the biofilm

The biofilm begins to be generated when an individual cell initially joins a surface. The ability of this cell to carry out this binding and initiate cell growth depends on factors such as temperature and pH of the medium, genetic factors that encode its motor functions and environmental sensitivity, and those that generate adhesins and other proteins [30–33]. The factors that directly affect its development depend mainly on the microbial species, although a part of these characteristics is common to numerous bacteria.

Once initial biofouling adhesion is produced, cell growth and expansion begin on the surface, forming monolayer microcolonies. At the same time, the cells modify their activity and begin the complex process of structure formation of the biofilm. The most obvious of these changes is the production of the exopolymer matrix (EPS), which will unite the whole [34]. If the conditions of the medium allow it, the biofilm will grow and spread to non-colonized areas releasing cells that will be distributed through the water in search of new surfaces to colonize.

The formation of the biofilm is a systematic process of predictable evolution, in which five phases are differentiated [35]: (1) the reversible adsorption of the bacterium to the surface, (2) the irreversible union, (3) the first maturation phase with growth and division, (4) the growth phase with production of the exopolymer, and (5) the final development of the colony with dispersion of colonizing cells (**Figure 8**).

3.2.2 Conditioning of the surface

The ability to bind to different materials depends on the specific proteins of its coat and the bacterial motor appendages. According to Pedersen [36], stainless steel can be as susceptible to the formation of biofilms as plastic. This is because the organic matter present in the water previously comes in contact with the surface, depositing an organic layer in the water/surface interface that changes the chemical and physical properties of the surface, improving the possibilities of fixing the bacteria.

3.2.3 Adsorption of molecules

The adsorption of ions and other dissolved substances (sugars, amino acids, proteins, fatty acids, etc.) begins when the material submerges under seawater

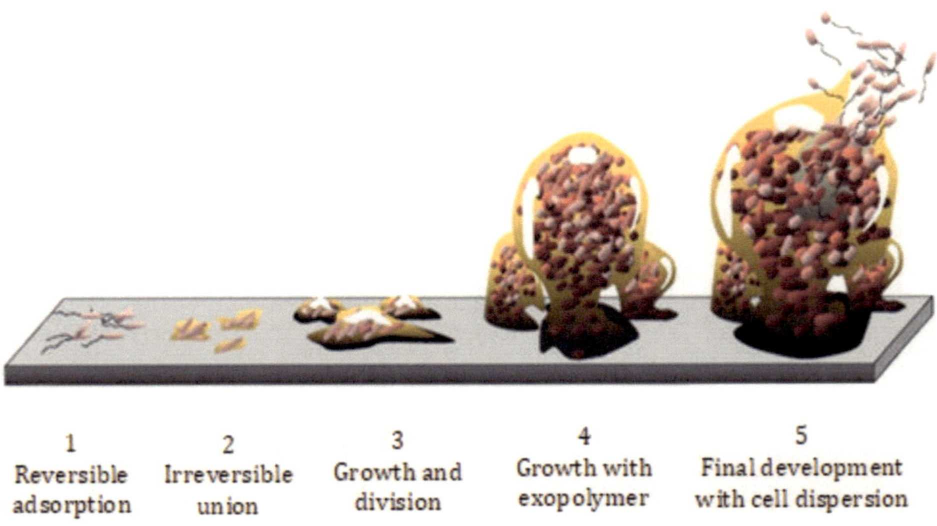

Figure 8.
Phases of biofilm formation (Source: By courtesy of Center Biofouling Engineering).

and accelerates when the fluid begins to circulate through the heat exchanger. This adsorption occurs rapidly on the surface, saturating the concentration of substances on the surface in a few minutes [37].

3.2.4 Adhesion of microorganisms

The adhesion of microorganisms to a substrate can be active (by flagella, fimbrias, adhesins, capsules, and surface charges) or passive (by gravity, diffusion, and fluid dynamics). In the absence of these mechanisms, the bacterial cells would be repelled by the surface when presenting electric charges of the same sign [38].

In a few minutes, the free bacteria form a reversible EPS matrix with the "conditioned" surface (**Figure 9**) [39], whose characteristics depend on the electrical charges of the bacteria. These attractive forces have their origin in hydrogen bonds, cation bonds, and van der Waals forces that compete with the forces of repulsion. If this union is maintained long enough, new chemical and physical structures appear that make it permanent and irreversible [40].

In cases of high microbial population density or lack of nutrients in the water, some microorganisms are able to individually alter their cell wall to make it more hydrophobic and increase its greater affinity of adhesion toward the surfaces. When the microorganisms approach the surface, with almost no water flow, they are attracted, proving their affinity for union and fixation (**Figure 10**) [41].

During the reversible adsorption stage, the bacterial cells still show Brownian motion and are easily removed with a nonaggressive cleaning method. The irreversible union implies the anchoring of bacterial appendages and the production of exopolymers, which determines that the mechanical action necessary to detach them will be greater depending on the time that the biofilm is active.

Bacteria undergo important transformations in their structure to adapt to the environment. These transformations activate different genes that encode new structural proteins and enzymes, which explains the adhesion and resistance of biofilm bacteria to antibiotics and disinfectants. In recent years, the advances made in the field of proteomics and genomics have allowed the identification of 800 proteins that modify their concentration throughout the five phases of the biofilm development and clarify the complex process of biofilm formation [42, 43].

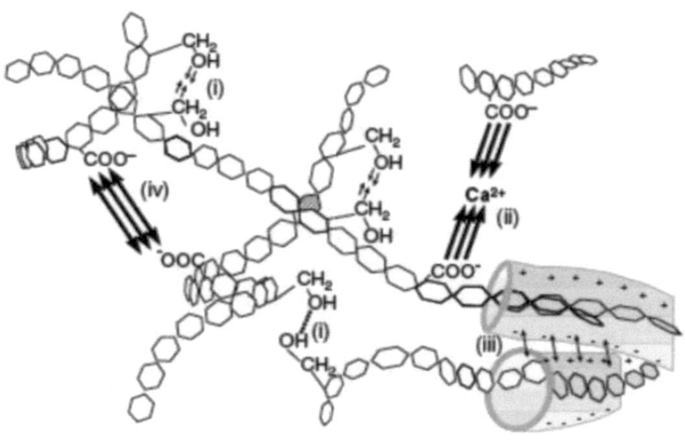

Figure 9.
Binding forces in an EPS matrix: (i) hydrogen bonds, (ii) cation bonds, (iii) van der Waals forces and (iv) repulsion force [38].

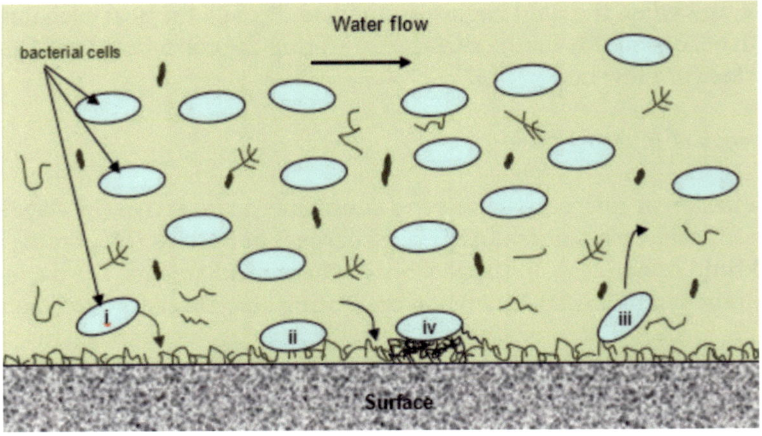

Figure 10.
Water flow transport of a bacterial cell to the conditioned surface: (i) adsorption, (ii) reversible adsorption, (iii) detachment and (iv) irreversible adsorption [40].

3.2.5 Maturation

Environmental stability favors biofouling growth and multiplication of cells which allows to generate a polyanionic polymer mixture of silty and sticky consistency whose are excreted to the outside to facilitate union cells onto surface.

Although its composition is not completely known, the mixture of exopolymers is formed by polysaccharides or glycoproteins of various sugars (glucose, fructose, mannose, N-acetylglucosamine, and others) [44] and, additionally, may contain free proteins, phospholipids, and nucleic or teichoic acids [45]. These free proteins are useful for retaining nutrients and protecting bacteria from various biocides.

Extracellular exopolymeric material or glycocalyx is expelled from the bacterial cell wall and adopts a reticular structure reminiscent of a spider. This structure is formed from groups of polysaccharides both neutral and carriers of electric charges, which act as ion exchange systems, capable of capturing and concentrating the nutrients present in the medium.

The structure of the biofilm interacts in a complex way, showing a behavior similar to that of multicellular organisms. If a microorganism generates toxic waste, another will use it as food and, in this way, coordinate the biochemical resources of all the beings that inhabit the matrix of the biofilm. In addition, some bacteria clump together within the matrix with a series of enzymes that allow them to digest nutrients that no isolated species could digest. Also, these enzymes will be used to respond to the attack of various biocides.

Anaerobic biofilms can be developed under the aerobic layer whose structure is permeabilized with a mesh of furrows crossed by water, bacterial debris, enzymes, nutrients, metabolites and oxygen. The gradients of ions and molecules that are established between the different zones generate the necessary impulse to take the substances to the periphery of the biofilm where most of the cells are located. The number of these cells is reduced with the age of the biofilm, being 80% in a young matrix and 50% in an aged shade [42].

3.2.6 Growth and dispersion

Biofilm formation has continuous divisions of the cellular matrix colony, which means a periodic detachment of groups of cells that deposit downstream.

These new colonies take advantage of the release of residues and nutrients from the original colonies to prepare the new surface and to feed other cells, and, as a consequence, the new colonies will grow and spread much more rapidly than in the original biofilm. This colonization is related to the long-term evolution and survival of bacteria [30].

If the conditions of the fluid allow it, the equilibrium that is established between the growth of the colony and the movement of the water releases few cells (**Figure 11**). With an intense or turbulent flow, many more can be released, even releasing whole areas of the biofilm colony.

3.2.7 Contribution of nutrients to the biofilm

The main factor controlling the growth of the biofilm is the availability of dissolved nutrients and their conversion into accumulated biomass. In cooling water circulation systems, the transfer of nutrients to the biofilm tends to increase with flow velocity [41]. Also, the rough surfaces of biofilms increase the transfer of nutrients about three times in relation to smooth surfaces [30, 46]. The control of nutrients is a way to control the development of the biofilm [47–49]. Melo and Bott [50] observed, in an industrial refrigeration system, an increase of 400% in the thickness of the biofilm at a speed of 1.2 ms^{-1} for an increase in the nutrient level of 4 mg L^{-1} at 10 mg L^{-1}.

The chemical composition of the waters determines the number, diversity, metabolic state of the bacteria, and their tendency to adhere to surfaces [51]. So far there has been no study that directly relates nutrients, biofilms, and colonization. Huang et al. [52] demonstrated in the laboratory that the availability of nutrients and the synthesis of new proteins for the formation of biofilms of *Pseudoalteromonas spongiae* under static conditions and without added nutrients affected the induction and adhesion of the biofilm to the surface. The effects of organic substances in the form of amino acids on the bioactivity of the biofilm were studied by Jin and Qian [53, 54]. The results of this study showed that the incorporation of aspartic acid and glutamic acid causes a significant increase in the bacterial mass, modifies its structure, and increases the inducing effect of biofilm formation. In addition, Huang et al. [55] found that the characteristics of biofilms generated in habitats with different environmental conditions show remarkable differences in the bioactivity of the larval settlements of the barnacle, which

Figure 11.
Microbial cell transport during the growth phases of the biofilm (Source: By courtesy of Center Biofouling Engineering).

suggests that the nature of the nutrients is a determining factor of the biological activity in the biofilm.

The nutrients required by microorganisms for feeding are divided into two broad categories [56]:

- Macronutrients (C, H, O, N, P, S, K, Mg)
- Micronutrients or trace elements (Co, Cu, Zn, Mo, etc.)

In nature, these elements are combined as part of organic and inorganic substances. Some of the nutrients will be incorporated to build macromolecules and cellular structures; others only serve for energy production not directly incorporated as cellular material, and others can perform both functions [56].

The bacterial colony exhibits a great metabolic versatility in the use of nutrients. Autotroph bacterials obtain their carbon by reducing CO_2 and other elements from inorganic sources. On the other hand, heterotroph bacterials use to wide ranges of organic carbon sources. In turn, within the heterotrophs, you can find many and varied types of nutrition, from methylotrophic bacteria that only use methane or methanol as carbon and energy sources to the very versatile *Pseudomonas*, which can resort to degrade more than 100 types of carbon sources (including aliphatic and cyclic hydrocarbons). In any case, among the heterotrophs, the most widespread source of carbon is glucose [56].

3.2.8 Abiotic and biotic factors that influence the development of the biofilm

Alterations in the structure of biofilms can be caused by abiotic factors such as [30]:

- Depth.
- Illumination.
- Exhibition time.
- Height of the tides.
- Water circulation regime.
- Physical alteration.
- Latitude.
- Season.
- Water chemistry.
- Supply of nutrients.
- Physical and chemical (abiotic) conditions at the seawater/biofilm interface are an important factor for the development of the biofilm. As the thickness of the biofilm increases, variations in pH, dissolved oxygen, and metabolic by-products are more important and generally directly influence its development [57].

The biotic factors that can alter the structure of the biofilm are [40]:

- Availability and physiological state of organisms, interaction between organisms, and biological disturbances [58–60]
- Properties of adhesion of microorganisms under conditions of turbulence.
- Physiological state of the organisms that have a selective advantage for the formation and development of the biofilm [61]. This capacity is reinforced by the secretion of EPS that are more resistant to high flow rates and chemical agents [62]. Microorganisms in the biofilm are released when conditions become unfavorable [63].

3.3 Effects produced by biofouling in heat exchangers

Biofouling has a great influence on the thermal performance of heat exchangers, due to the accumulation of biotic deposits with insulating characteristics in heat exchange surfaces. This accumulation adds not only an additional thermal resistance to the flow of heat but a greater frictional resistance to the passage of the fluid. The formation of these biofilms produces important modifications, since they alter the physical–chemical conditions at the metal-solution interface and form barriers for the exchange of elements between the metallic surface and the surrounding liquid medium [64]. Among the negative consequences, the decrease of the heat output of the heat exchangers and the lower durability of the construction materials of the equipment can be mentioned [65].

The main consequences produced by microfouling (biofouling) in heat exchangers are the following [5]:

- Production losses due to the decrease in efficiency and the scheduled and unscheduled shutdowns of the installation
- Maintenance costs, resulting from the elimination of biofouling deposits with chemical and/or mechanical devices
- Increased corrosion processes in metallic components, induced by microbiological corrosion (MIC) [66]
- Increased consumption of water, electricity, fossil fuels, and other sources to counteract the effects of biological fouling
- Increase in manufacturing costs of heat exchangers to consider an acceptable biological fouling without losing design power and to be able to mechanically clean them
- Increase in environmental risks due to the use of biocides and CO_2 emissions to the environment to counteract biological fouling

The main consequences of macrofouling in a heat exchanger are [40]:

- High costs caused by the complexity of the cleaning of the macroorganisms of the installation
- Losses of power produced by the loss of flow and thermal efficiency in the heat exchangers of a refrigeration installation

- Costs produced by the management of the macrofouling waste after its removal from the installation
- Production of faults due to the accumulation of macroorganisms in auxiliary systems, such as valve drives, grids, pumps, instrumentation, etc.

4. Design aspects that minimize the adhesion of biofouling

The design of the heat exchanger can represent an important factor to minimize the presence of biofouling on its surfaces. The main design variables that intervene in the biofouling growth process are [27]:

- Speed of the fluid
- Composition and roughness of the support
- Temperature of the support and cooling water
- Physicochemical properties of the fluid
- Biological variables
- Geometry of fluid passage

Although it is very difficult to relate the colonization of the surface with only one of these variables, it can be affirmed that they all have their influence on the process of developing the biofilm on the surface in contact with seawater. In this way, it is recommended to maintain minimum flow rates above 0.9 ms^{-1} and use smooth surfaces and high temperatures (>90°C). Otherwise, the presence of biofouling on the heat exchange surface is unavoidable.

5. Chemical antifouling treatments

The chemical compounds used as antifouling agents in heat exchangers are called biocides. In addition, sequestrants, dispersants, or chelators can be used for other purposes (**Table 1**).

The biocides contain one or more active substances, whose function is to destroy, counteract, neutralize, and impede the action or exert a control of another type on any harmful organism by chemical or biological means. Properly dosed reduce drastically the number and activity of cells and biofilms in water. Due to its toxic nature, inadequate doses can be harmful to the environment and are therefore

Agent	Methodology
Sequester	Form a chemical complex with the tank keeping it in suspension
Dispersant	It transmits load to the surface keeping the particles in suspension
Chelator	It forms a strong chemical complex with the species that form the encrustation
Biocide	Kills or incapacitates generating irreversible damage to the structure or vital functions of microorganisms

Table 1.
Chemical agents used in the treatment of seawater [56].

Biocide	Ionic character	pH	Optimal activity
Cl_2 gas	No	5–7.5	Algae and bacteria
NaClO	Anionic	7–8.5	Algae and bacterias
Br_2	No	5–9	Algae and bacteria
ClO_2	No	6–10	Algae, bacteria, fungi, and sulfate reducing bacteria
Chlorophenols	Anionic	5–8	Algae, bacteria, and fungi
Sales de NH_4^+	Cationic	7–9.5	Algae and bacteria
Sulfur organ	Anionic	5–7.5	Bacteria and fungi
Brominated organ	No	5–8	Algae, bacteria, fungi and sulfate reducing bacteria
Isothiazolones	No	5–8.5	Algae, bacteria, fungi, and sulfate reducing bacteria
Glutaraldehyde	No	5–8.5	Algae, bacteria, fungi, and sulfate reducing bacteria

Table 2.
Specific operating conditions of different biocides [67].

Common name	Compound
BNP	2-Bromo-2-nitro propane-1,3-diol
BNS	Beta-bromo-beta-nitrostyrene
Carbamates	Sodium dimethyl dithiocarbamate, ethylene bis dithiocarbamate disodium, disodium cyanodithioimidocarbamate, Potassium N-methyldithiocarbamate, potassium methyldithiocarbamate
Chlorine	Sodium hypochlorite, calcium hypochlorite trichloroisocyanuric acid, sodium dichloroisocyanide and potassium acid dichloroisocyanide
Chlorophenols	Sodium pentachlorophenate and sodium trichlorophenate
Dazomet	Tetrahydro-3,5-dimethyl-1-2-H-1,3,5-thiadiazine-2-thione
DBNPA	2,2-Dibromo-3-nitrile-propiomide
DGH	Dodecylguanidine hydrochloride
Isothiazolone	5-Chloro-2-methyl-4-isothiazolon-3-one
Glutaraldehyde	1,5-Pentadial
HPMTS	Hydroxypropyl methanethiosulfonate
Hydantoin	1-Bromo-3-chloro-5,5-dimethylhydantoin, 1,3 dichloro-5,5 dimethyl hydantoin, 1,3-dichloro-5-ethyl-5-ethylhydantoin
MBT	Methylene-bis- (thiocyanate)
Sodium bromide	Sodium bromide
Sulfone	Bis- (trichloromethyl) -sulfone
TBTO	Bis- (tri-n-butyl) tin oxide
TCMTB	2(Tiocianatometiltio) benzotiazol
Triacin	Dimethyl propyl-n-ethyl-6- (methylthio) triazine diamine Chloro-4-ethylamino-6-t-butylamino-s-triazine
Tris-nitro	Tris (hydroxymethyl) nitro methane
Quaternary ammonium chloride	Didecyl dimethyl ammonium chloride, n-alkyl dimethyl benzyl ammonium chloride, dialkyl dimethyl ammonium chloride poly oxyethylene (dimethyl ammonium), ethylene (dimethylammonium) ethylene dichloride

Table 3.
Biocides used in cooling water [71].

subject to strict regulations of use [10]. These rules are increasingly restrictive, so that the situation can be reached and that the concentration of allowed biocide is lower than the effective, which requires the search for other alternatives.

Biocides used in industrial refrigeration systems must have a broad spectrum, as they must limit the growth of a wide variety of microorganisms including bacteria, fungi, and algae. In addition, they must be effective in a wide range of specific conditions of equipment operation (**Table 2**).

There is a wide variety of chemical products that have been used as biocides to inhibit the formation of biofouling deposits in heat exchange systems that use seawater as a cooling fluid. According to their general principle of action, they are classified as oxidants or as non-oxidants.

They act by oxidizing the organic matter (cellular matter, enzymes, proteins, etc.) of the microorganism, and as a consequence they cause its death [68]. Its use is widespread, and its effectiveness as a widely demonstrated antifouling agent [69, 70].

In industrial refrigeration systems, different chemical compounds are employed that have the function of drastically reducing the number of cells present in the seawater-surface interface of heat exchange and minimizing their activity, having the capacity to limit the growth of a wide variety of microorganisms (bacteria and algae) and to be effective in a wide range of equipment operating conditions (**Table 3**).

6. Biological treatments antifouling

Biological antifouling biocides are active principles created by the agencies themselves (algae, batteries, fungi, etc.) as self-defense against other organisms in their natural environment. These substances are usually of a protein nature and are called enzymes.

Biological biocides are in the initial phase of research and development. Its use requires an in-depth analysis of the organisms that are intended to be eliminated and the environment in which they are found. The most recent research has shown the inhibition of marine biofouling by different types of bacteria, cyanobacteria, and marine fungi. Recent experiments have shown the antifouling properties of some organic compounds [72, 73], marine larvae of epibiotic bacteria [74–76], and various natural chemical compounds [77]. Dobretsov et al. [78] studied the microbial interference as a mechanism to control marine biofilms, focusing on the mechanisms of altering the biofilm by bacteria, algae, and various larvae.

7. Physical antifouling treatments

They represent an alternative to the use of biocides and an emerging research channel. They can be applied with the installation in operation (online systems) or with the installation in technical stop (offline systems) [5].

7.1 Online systems

7.1.1 Taprogge system

Mechanical cleaning system by sponge balls or elastomer material abrasives is a biofouling treatment which periodically introduce balls by the water flow to clean tubes in heat exchangers. The balls of suitable roughness are introduced into the water box of the condenser from where they pass into the tubes. Balls have a diameter greater than that of the tube which pass through tubes to clean inner surface avoiding the settlement of organic or inorganic matter.

7.1.2 Cleaning brush system housed in baskets

At the ends of the tubes, there are baskets containing a brush that moves in both directions through the interior of the tube by reversing the flow of the cooling water. The investment of flow is made with a valve system that enables this action, not being necessary more than three daily cycles to maintain good conditions of the plant.

7.1.3 Heat treatment

It consists of heating the cooling water by recirculating it in a closed circuit without renewing cold water or by installing a secondary hot water supply circuit from an internal combustion engine, for example. The water recirculates until reaching temperatures between 50 and 70°C, which produces the death of microorganisms [5]. The periodicity of this treatment is established taking into account the biological activity of the water. Its effectiveness depends mainly on the temperature of the water, the time of exposure, and the frequency of repetition [40].

7.1.4 Variation in cooling water flow rate

The speed of water flow through the tubes of the exchanger is a determining factor in the growth of biofouling. In addition to influencing the contribution of nutrients that determine the growth of organic matter that is part of the biofouling film, deposits adhered to the inner surface of the tubes can be detached as a result of the greater cutting effect on the surface are exerted to higher water flow rates. The heat exchanger can be designed to work at high flow rates (from 1 to 2 ms^{-1}) or so that the water flow rate can be varied for a short period of time (e.g., from 1 to 3 ms^{-1}) at regular intervals. This increase in flow velocity is only possible if the pumping capacity of the system can be increased.

Pulsed injections can also be made by introducing a gas (compressed air or nitrogen) at very high speed (3 ms^{-1}) periodically at short intervals of time (5, 10, and 60 minutes). The effect is to increase the turbulent regime, creating pressure fluctuations inside the tube and higher shear stresses on the surface. As a consequence, a greater detachment of biofouling is obtained than that obtained by simply increasing the speed of water flow [5]. However, according to Mussalli and Tsou [79], the effectiveness of this system decreases as the biofouling film becomes more compact.

7.1.5 Streams current antifouling

The principle of operation of this system is based on the generation of a voltage between the copper anodes (Cu) and the integrated steel cathodic plates so that when the Cu dissolves, the medium is contaminated by repelling the marine organisms. The level of Cu dissolution is regulated by a control unit adapting the system to the factors that influence biofouling growth (e.g., the season of the year) and lengthening its useful life.

7.1.6 Ultraviolet radiation

The sterilization of water by ultraviolet radiation (UV) is a potentially useful method to avoid the formation of the biofouling film and as a cleaning technique. Its principle of operation is based on attacking the deoxyribonucleic acid (DNA) of the cells, causing their death. In this way, the amount of microorganisms present in the cooling water flow is minimized.

This treatment is very effective against microorganisms in the larval stage, and their environmental impact is zero [5]. On the contrary, the useful life of UV lamps is limited (approximately 8 months), and the cost of operation and maintenance is high, which is why it is not normally used in industrial installations that require a large flow of water to treat [79].

In order to reduce the treatment and broaden the spectrum of organisms attacked, it is usually combined with another chemical or physical method [5].

7.1.7 Ultrasound

These are acoustic waves whose mode of action is based on the creation of vibrations to generate cavitation bubbles. The size of the bubbles can be increased by means of compression-decompression cycles until reaching a critical value at which they explode transforming their energy into heat. These microexplosions produce a disturbance in the surface that breaks the cohesion forces of the deposits to the surface causing their detachment [5]. Inside the cavitation bubbles, the temperature and pressure conditions can reach 5000°C and 500–2500 bar [80, 81], so that even water molecules decompose generating HO- and H+ radicals. The radicals formed can be recombined in the same way or react with substances present in the medium causing their degradation. Obviously, the generation of radicals is facilitated if in the medium there are molecules that break easily, such as ozone or hydrogen peroxide, which also act as usual precursors of hydroxyl radicals in the oxidation processes. The greater the effectiveness of ultrasound treatments, the higher their frequency (100 Hz and 100 MHz).

This antifouling treatment is effective against microorganisms in the larval stage and also has no environmental impact. On the contrary, the high cost of the ultrasonic units must be added to the energy cost of the installation.

7.1.8 Filtering systems

These systems are considered complementary methods for the elimination of biofouling in any industrial facility. The filtrate prevents larvae of mussels, crustaceans, and other organisms from entering the circuits of refrigeration systems and therefore develops into a biofouling film.

In cooling water conduction systems, the most commonly used filters are of the panel type for flow rates of less than 10 $m^3\ s^{-1}$ and of the drum type for higher flow rates [40]. Before the installation, the overall cost/benefit of the system must be analyzed due to the pressure drop produced in the installation [79].

7.1.9 Injection of fresh water in seawater systems

The microorganisms present in the biofilm feed on the nutrients existing in the water through osmotic processes through a semipermeable membrane. When injecting fresh water, the saline concentration of the water is modified, and the pressure inside the cells increases, causing death [5]. According to Cho et al. [82], the exposure time must be high, reaching 48 hours for the mussel *Mytilus californianus* and 63 hours for the mussel *Mytilus edulis*. This method is used in piping systems of refrigeration systems that remain out of service for long periods [83].

7.1.10 Inserts in the tubes

This system consists of inserting a helical element, whose vibration, distortion of flow, or rotation mitigate the growth of biofouling adhering on the inner surface of the tubes [5].

This system can present the following variants [84]:

- SPIRELF®. Metallic helical element attached to the ends of the tube and adjusted to its length. When the fluid circulates inside the tube, it vibrates axially and radially, eliminating biofouling adhering to its walls.

- FIXOTAL® or HITRAN®. They use metallic helical elements that are fixed inside the tubes, improving the turbulent effect on the wall to reduce the thickness of biofouling.

- TURBOTAL®. Rigid helical device that is inserted into the tubes and uses the water flow itself as the impeller of the rotating cleaning mechanism.

The insertion of metal elements inside the tubes has proven to be an effective method to prevent the formation of biofouling in heat exchangers of different industrial processes [84]. However, the use of this type of device means increasing the cost of the equipment and its maintenance, as well as a higher consumption of pumping energy due to the increase of load losses in the system.

7.1.11 Circulation of polymers

The method consists of introducing polymer fibers entrained by the fluid stream in order to erode and detach the layer of adhered deposits. According to Bott [5], its effectiveness on the process of forming the biofouling film in the tubes of a heat exchanger depends on the concentration of fibers and the speed of the flow of cooling water. The biggest drawback is that of removing the fibers from the effluent before it is discharged into the natural environment.

7.1.12 Treatments with electromagnetic fields

Electromagnetic fields have the ability to increase the interaction of the Ca^{2+} and $CO3^{2}-$ ions present in seawater, increasing their nucleation and precipitation in $CaCO_3$. Below 35°C, $CaCO_3$ precipitates in the form of aragonite instead of calcite, resulting in being less adherent and less insulating [3, 85, 86]. The precipitation process of $CaCO_3$ reduces the superficial tension of the water, minimizing the adherence of microorganisms, and affects the intermolecular union of the extracellular polymers, weakening the matrix of the biofouling layer and diminishing its capacity of adhesion to the surface [7]. In addition, $CaCO_3$ entrainment by the seawater flow produces an erosive effect on the weakened biofouling layer that reduces its thickness [87, 88].

7.2 Offline systems

The use of online antifouling treatments does not prevent periodically scheduled stops in which aggressive cleaning methods are used to recover the efficiency of the tube at 100%. The frequency of scheduled stops is based on experience, and the cost/benefit ratio is derived from its application.

Brushing the surfaces of the heat exchanger in a dry or wet surface condition to eliminate adhering deposits. The cost of cleaning is high and involves access to the surfaces of the exchanger.

Although the techniques used to clean the surfaces of an industrial heat exchanger are varied, the most common methods to clean the internal surface of the pipes are air, water, and water steam.

The choice of cleaning agent depends on the degree of adhesion of the tanks and the design of the heat exchanger. According to Wilson [89], the use given to the different cleaning agents is as follows:

- Air under pressure. Eliminates deposits that are not very adherent
- Water at high pressure. Eliminates deposits and calcium deposits in tubular exchangers, condensers, and low-pressure boilers
- Water vapor. Eliminates calcareous incrustations and other substances of great hardness

To increase cleaning efficiency, detergents or abrasive materials are added to water or air under pressure. The nature of the abrasive materials depends on the hardness of the tanks and the construction material of the equipment. Sand is used as abrasive materials, pulverized mollusk shells, etc. [89].

8. Conclusions

Nowadays, heat exchangers play a vital role in the operational efficiency and effectiveness at every processing facility in which heat is essential, whether it is an oil refinery, combined heat and power plants, solar and biomass power plant, etc. As energy efficiency importance grows, the role of heat exchangers will become even greater and their technologies more advanced. Fouling on heat transfer surfaces of power plants are a major economic and environmental problem worldwide. Estimates have been made of fouling costs due primarily to wasted energy through excess fuel burn that are as high as 0.25% of the gross national product (GNP) of the industrialized countries. Many millions of tons of carbon emissions are the result of antifouling inefficiency.

After this study of the state of the art of fouling in heat exchanger, the main conclusion is that fouling problem is latent and unresolved. For this reason, new research projects must be developed to investigate and resolve this old problem without environmental effect.

Author details

Sergio García* and Alfredo Trueba
Biofouling Group Research, University of Cantabria, Santander, Spain

*Address all correspondence to: sergio.garcia@unican.es

References

[1] Wahl M. Marine epibiosis. I. Fouling and antifouling: Some basic aspects Marine Ecology Progress Series. 1989;**58**:175-189. https://doi.org/10.3354/meps058175

[2] Shah RK, Sekulić DP. Fundamentals of Heat Exchanger Design. John Wiley & Sons. 2003. 941 p. DOI: 10.1007/BF00740254

[3] Trueba A, García S, Otero FM. Mitigation of biofouling using electromagnetic fields in tubular heat exchangers-condensers cooled by seawater. Biofouling. 2014;**30**:95-103. DOI: 10.1080/08927014.2013.847926

[4] Trueba A, Otero FM, González JA, Vega LM, García S. Study of the activity of quaternary ammonium compounds in the mitigation of biofouling in heat exchangers–condensers cooled by seawater. Biofouling. 2013;**29**:1139-1151. DOI: 10.1080/08927014.2013.830108

[5] Bott TR. Industrial Biofouling. Elsevier. 2011. 220 p. ISBN: 9780444532244

[6] Paul VJ, Ritson-Williams R, Sharp K. ChemInform abstract: Marine chemical ecology in benthic environments. ChemInform. 2011;**28**(2):345-387. DOI: 10.1002/chin.201121256

[7] Trueba A, García S, Otero FM, Vega LM, Madariaga E. The effect of electromagnetic fields on biofouling in a heat exchange system using seawater. Biofouling. 2015;**31**:19-26. DOI: 10.1080/08927014.2014.994096

[8] García S, Trueba A. Influence of the reynolds number on the thermal effectiveness of tubular heat exchanger subjected to electromagnetic field-based antifouling treatment in an open once-through seawater cooling system. Applied Thermal Engineering. 2018;**140**:531-541. DOI: 10.1016/j.applthermaleng.2018.05.069

[9] García S, Trueba A, Vega LM, Madariaga E. Ceramic coating solution for offshore structures. Progress in Maritime Technology and Engineering. 2018;**1**:453-460. ISBN: 9781138585393

[10] Eguía E, Trueba A. Application of marine biotechnology in the production of natural biocides for testing on environmentally innocuous antifouling coatings. Journal of Coatings Technology and Research. 2007;**4**:191-202. DOI: 10.1007/s11998-007-9022-3

[11] Eguía López E. El problema del biofouling en intercambiadores de Calor-condensadores refrigerados por agua de mar. Universidad de Cantabria. 1998. 157 p

[12] Somerscales EFC, Kassemi M. Fouling due to corrosion products formed on a heat transfer surface. Journal of Heat Transfer. 1987;**109**:267-271. DOI: 10.1115/1.3248061

[13] Huang JY, John Chew YM, Ian Wilson D. A spinning disc study of fouling of cold heat transfer surfaces by gel formation from model food fat solutions. Journal of Food Engineering. 2012;**109**:49-61. DOI: 10.1016/j.jfoodeng.2011.09.034

[14] Bott TR. Freezing fouling or liquid solidification. Fouling of Heat Exchangers. Elsevier. 1995:137-147. 546 p. ISBN: 0080531903

[15] Garcia S, Trueba A, Vega L, Madariaga E. Improvement of electromagnetic fields treatment for biofouling growth control in tubular heat exchanger-condenser cooled by seawater. In: Ocean 2017 - Aberdeen. IEEE. 2017. DOI: 10.1109/OCEANSE.2017.8084576

[16] Boerlage SFE, Kennedy MD, Bonne PAC, Galjaard G, Schippers JC. Prediction of flux decline in membrane systems due to particulate fouling. Desalination. 1997;**113**:231-233. DOI: 10.1016/S0011-9164(97)00134-3

[17] Eguía E, Trueba A, Río-Calonge B, Girón A, Bielva C. Biofilm control in tubular heat exchangers refrigerated by seawater using flow inversion physical treatment. International Biodeterioration and Biodegradation. 2008;**62**:79-87. DOI: 10.1016/j.ibiod.2007.12.004

[18] Berkey K, BenDor TK. A comprehensive solution to the biofouling problem for the endangered Florida manatee and other species. Journal of Environmental Law. 2012;**42**:415-468

[19] Callow JA, Callow ME. Advanced nanostructured surfaces for the control of marine biofouling: The AMBIO project. In: Advances in Marine Antifouling Coatings and Technologies. Woodhead Publishing Series in Metals and Surface Engineering. 2009. p. 647-663. DOI: 10.1533/9781845696313.4.647

[20] Satpathy KK, Mohanty AK, Sahu G, Biswas S, Selvanayagam M. Biofouling and its control in seawater cooled power plant cooling water system—A review. In: Nuclear Power. IntechOpen; 2010. pp. 191-243. DOI: 10.5772/9912. ISBN: 978-953-307-110-7

[21] Holm ER. Barnacles and biofouling. Integrative and Comparative Biology. 2012;**52**:348-355. DOI: 10.1093/icb/ics042

[22] García S, Trueba A, Vega LM, Madariaga E. Impact of the surface roughness of AISI 316L stainless steel on biofilm adhesion in a seawater-cooled tubular heat exchanger-condenser. Biofouling. 2016;**32**:1-9. DOI: 10.1080/08927014.2016.1241875

[23] Zobell CE. The effect of solid surfaces upon bacterial activity. Journal of Bacteriology. 1943;**8**:1-8. DOI: 10.1146/annurev.arplant.58.032806.103848

[24] Jones HC, Roth IL, Sanders WM. Electron microscopic study of a slime layer. Journal of Bacteriology. 1969;**99**:316-325

[25] Costerton JW, Geesey GG, Cheng KJ. How bacteria stick. Scientific American. 1978;**238**:86-95. DOI: 10.1038/scientificamerican0178-86

[26] Flemming HC, Wingender J. Biofilms as a particular form of microbial life. Biofouling and Biocorrosion in Industrial Water Systems. Springer. 1991. DOI: 10.1007/978-3-642-76543-8

[27] Cloete TE, Brözel VS, Von Holy A. Practical aspects of biofouling control in industrial water systems. International Biodeterioration and Biodegradation. Elsevier. 1992;**29**:299-341. DOI: 10.1016/0964-8305(92)90050-X

[28] Mah TFC, O'Toole GA. Mechanisms of biofilm resistance to antimicrobial agents. Trends in Microbiology. 2001;**9**:34-39. DOI: 10.1016/S0966-842X(00)01913-2

[29] Flemming HC, Sriyutha Murthy P, Venkatesan R, Cooksey K. Marine and Industrial Biofouling. Spinger. 2008. 334 p. DOI: 10.1007/978-3-540-69796-1

[30] Characklis WG, Marshall KC, editors. Biofilms: Basis for an Interdisciplinary Approach. New York: Wiley; 1990. pp. 3-15. DOI: 10.1081/E-EBBE-120007264

[31] Costerton JW. Overview of microbial biofilms. Journal of Industrial Microbiology. 1995;**15**:137-140. DOI: 10.1007/BF01569816

[32] O'Toole G, Kaplan HB, Kolter R. Biofilm formation as microbial development. Annual Review of

Microbiology. 2000;**54**:49-79. DOI: 10.1146/annurev.micro.54.1.49

[33] Kraigsley AM, Finkel SE. Adaptive evolution in single species bacterial biofilms. FEMS Microbiology Letters. 2009;**293**:135-140. DOI: 10.1111/j.1574-6968.2009.01526.x

[34] Danese PN, Pratt LA, Kolter R. Biofilm formation as a developmental process. Methods in Enzymology. 2001;**336**:19-26. DOI: 10.1016/s0076-6879(01)36574-6

[35] McFeters GA, Bazin MJ, Mirelman D, Bryers JD, Mitchell R, Caldwell DE, et al. Biofilm development and its consequences. In: Report of the Dahlem Workshop on Microbial Adhesion and Aggregation; Berlin: Springer; 15-20 January 1984. p. 108-124. DOI: 10.1007/978-3-642-70137-5_9

[36] Pedersen K. Biofilm development on stainless steel and PVC surfaces in drinking water. Water Research. 1990;**24**:239-243. DOI: 10.1016/0043-1354(90)90109-J

[37] Railkin AI, Ganf TA, Manylov OG. Marine biofouling colonization processes and defenses. Biofouling. 2004;**20**:129-131. DOI: 10.1038/467531c

[38] Boonaert CJP, Dufrene YF, Rouxhet PG. Adhesion (primary) of microorganisms onto surfaces. In: Encyclopedia Environmental Microbiology. John Wiley & Sons. 2003. DOI: 10.1002/0471263397.env154. ISBN: 9780471354505

[39] Mayer C, Moritz R, Kirschner C, Borchard W, Maibaum R, Wingender J, et al. The role of intermolecular interactions: Studies on model systems for bacterial biofilms. International Journal of Biological Macromolecules. 1999;**26**:3-16. DOI: 10.1016/S0141-8130(99)00057-4

[40] Flemming H. Why microorganisms live in biofilms and the problem of biofouling. Marine and Industrial Biofouling. 2009;**4**:3-12. DOI: 10.1007/7142

[41] Characklis WG, Marshall K. Biofilms: A basis for an interdisciplinary approach. Biofilms. John Wiley & Sons. 1990

[42] Whiteley M, Ott JR, Weaver EA, McLean RJC. Effects of community composition and growth rate on aquifer biofilm bacteria and their susceptibility to betadine disinfection. Environmental Microbiology. 2001;**3**:43-52. DOI: 10.1046/j.1462-2920.2001.00158.x

[43] Singh PK, Parsek MR, Greenberg EP, Welsh MJ. A component of innate immunity prevents bacterial biofilm development. Nature. 2002;**417**:552-555. DOI: 10.1038/417552a

[44] Parikh A, Madamwar D. Partial characterization of extracellular polysaccharides from cyanobacteria. Bioresource Technology. 2006;**97**:1822-1827. DOI: 10.1016/j.biortech.2005.09.008

[45] Chmielewski RAN, Frank JF. A predictive model for heat inactivation of listeria monocytogenes biofilm on stainless steel. Journal of Food Protection. 2016. DOI: 10.4315/0362-028x-67.12.2712

[46] Lazarova V, Manem J. Biofilm characterization and activity analysis in water and wastewater treatment. Water Research. 1995;**29**:2227-2245. DOI: 10.1016/0043-1354(95)00054-O

[47] Chandy JP, Angles ML. Determination of nutrients limiting biofilm formation and the subsequent impact on disinfectant decay. Water Research. 2001;**35**:2677-2682. DOI: 10.1016/S0043-1354(00)00572-8

[48] Griebe T, Flemming HC. Biocide-free antifouling strategy to protect RO membranes from biofouling. Desalination. 1998;**118**:153-156. DOI: 10.1016/S0011-9164(98)00113-1

[49] Flemming HC, Wingender J. Extracellular polymeric substances: Structure, ecological functions, technical relevance. Encyclopedia of Environmental Microbiology. John Wiley & Sons. 2002

[50] Melo LF, Bott TR. Biofouling in water systems. Experimental Thermal and Fluid Science. 1997;**14**:375-381. DOI: 10.1016/S0894-1777(96)00139-2

[51] Schneider RP, Marshall KC. Retention of the Gramnegative marine bacterium SW8 on surfaces—Effects of microbial physiology, substratum nature and conditioning films. Colloids and Surfaces. B, Biointerfaces. 1994;**2**:387-396. DOI: 10.1016/0927-7765(94)80002-2

[52] Huang YL, Dobretsov S, Xiong H, Qian PY. Effect of biofilm formation by Pseudoalteromonas spongiae on induction of larval settlement of the polychaete Hydroides elegans. Applied and Environmental Microbiology. 2007;**73**:6284-6288. DOI: 10.1128/AEM.00578-07

[53] Jin T, Qian PY. Effect of amino acids on larval metamorphosis of the polychaete Hydroides elegans. Marine Ecology Progress Series. 2004;**267**:209-218. DOI: 10.3354/meps267209

[54] Jin T, Qian PY. Amino acid exposure modulates the bioactivity of biofilms for larval settlement of Hydroides elegans by altering bacterial community components. Marine Ecology Progress Series. 2005;**297**:169-179. DOI: 10.3354/meps297169

[55] Huang YL, Dobretsov S, Ki JS, Yang LH, Qian PY. Presence of acyl-homoserine lactone in subtidal biofilm and the implication in larval behavioral response in the polychaete Hydroides elegans. Microbial Ecology. 2007;**54**:384-392. DOI: 10.1007/s00248-007-9210-9

[56] Bott TR. Fouling of Heat Exchangers. Elsevier. 1995. DOI: 10.1016/B978-044482186-7/50005-1

[57] Wimpenny J, Manz W, Szewzyk U. Heterogeneity in biofilms. FEMS Microbiology Reviews. 2000;**24**:661-671. DOI: 10.1016/S0168-6445(00)00052-8

[58] Clare AS, Rittschof D, Gerhart DJ, Maki JS. Molecular approaches to nontoxic antifouling. Invertebrate Reproduction and Development. 1992;**22**:67-76. DOI: 10.1080/07924259.1992.9672258

[59] Dahms HU, Qian PY. Exposure of biofilms to meiofaunal copepods affects the larval settlement of Hydroides elegans (Polychaeta). Marine Ecology Progress Series. 2005;**297**:203-214. DOI: 10.3354/meps297203

[60] Fenchel T. Formation of laminated cyanobacterial mats in the absence of benthic fauna. Aquatic Microbial Ecology. 1998;**14**:235-240. DOI: 10.3354/ame014235

[61] Beech IB, Gubner R, Zinkevich V, Hanjangsit L, Avci R. Characterisation of conditioning layers formed by exopolymeric substances of pseudomonas NCIMB 2021 on surfaces of AISI 316 stainless steel. Biofouling. 2000;**16**:93-104. DOI: 10.1080/08927010009378435

[62] Stewart PS. Mechanisms of antibiotic resistance in bacterial biofilms. International Journal of Medical Microbiology. 2002;**292**: 107-113. DOI: 10.1078/1438-4221-00196

[63] Piola RF, Dafforn KA, Johnston EL. The influence of antifouling practices on marine invasions. Biofouling. 2009;**25**:633-644. DOI: 10.1080/08927010903063065

[64] Beech IB, Sunner JA, Hiraoka K. Microbe-surface interactions in

biofouling and biocorrosion processes. International Microbiology. 2005;**8**:157-168. PMID: 16200494

[65] Caputo AC, Pelagagge PM, Salini P. Joint economic optimization of heat exchanger design and maintenance policy. Applied Thermal Engineering. 2011;**31**:1381-1392. DOI: 10.1016/j.applthermaleng.2010.12.033

[66] Coetser SE, Cloete TE. Biofouling and biocorrosion in industrial water systems. Critical Reviews in Microbiology. 2005;**31**:213-232. DOI: 10.1080/10408410500304074

[67] Barry DM, Kanematsu H. Cooling water. In: Biofilm and Materials Science. Springer. 2015. DOI: 10.1007/978-3-319-14565-5_10

[68] Nwaoha C, Zakaria ZY. Cooling towers. In: Process Plant Equipment: Operation, Control, and Reliability. John Wiley & Sons. 2012. DOI: 10.1002/9781118162569.ch5

[69] Fisher DJ, Burton DT, Yonkos LT, Turley SD, Ziegler GP. The relative acute toxicity of continuous and intermittent exposures of chlorine and bromine to aquatic organisms in the presence and absence of ammonia. Water Research. 1999;**33**:760-768. DOI: 10.1016/S0043-1354(98)00278-4

[70] Murphy HM, Payne SJ, Gagnon GA. Sequential UV- and chlorine-based disinfection to mitigate Escherichia coli in drinking water biofilms. Water Research. 2008;**42**:2083-2092. DOI: 10.1016/j.watres.2007.12.020

[71] Dixon D, Veil J, Wisniewski J. Defining and assessing adverse environmental impact from power plant impingement and entrainment of aquatic organisms. Symposium in Conjunction with the Annual Meeting of the American Fisheries Society, 2001. In Phoenix, Arizona, USA: CRC Press; 2003. 292 p. ISBN: 9789058095176

[72] Qian PY, Xu Y, Fusetani N. Natural products as antifouling compounds: Recent progress and future perspectives. Biofouling. 2009;**26**:223-234. DOI: 10.1080/08927010903470815

[73] Fusetani N. Antifouling marine natural products. Natural Product Reports. 2011;**28**:400-410. DOI: 10.1039/c0np00034e

[74] Hadfield MG. Biofilms and marine invertebrate larvae: What bacteria produce that larvae use to choose settlement sites. Annual Review of Marine Science. 2010;**3**:453-470. DOI: 10.1146/annurev-marine-120709-142753

[75] Thiyagarajan V. A review on the role of chemical cues in habitat selection by barnacles: New insights from larval proteomics. Journal of Experimental Marine Biology and Ecology. 2010;**392**:22-36. DOI: 10.1016/j.jembe.2010.04.030

[76] Wahl M, Goecke F, Labes A, Dobretsov S, Weinberger F. The second skin: Ecological role of epibiotic biofilms on marine organisms. Frontiers in Microbiology. 2012;**23**:292-313. DOI: 10.3389/fmicb.2012.00292

[77] Puglisi MP, Sneed JM, Sharp KH, Ritson-Williams R, Paul VJ. Marine chemical ecology in benthic environments. Natural Product Reports. 2014;**31**:1510-1553. DOI: 10.1039/c4np00017j

[78] Dobretsov S, Abed RMM, Teplitski M. Mini-review: Inhibition of biofouling by marine microorganisms. Biofouling. 2013;**29**:423-441. DOI: 10.1080/08927014.2013.776042

[79] Mussalli YG, Tsou J. Advances in biofouling control technologies. U.S. and Japanese perspectives. Proceedings of the American Power Conference. Illinois-Institute of Technology. 1989;**1**:1094-1099

[80] Wu TY, Guo N, Teh CY, Hay JXW. Advances in Ultrasound Technology for Environmental Remediation. Springer. 2013. ISBN 978-94-007-5533-8

[81] Park JS, Lee JH. Sea-trial verification of ultrasonic antifouling control. Biofouling. 2018;**34**:98-110. DOI: 10.1080/08927014.2017.1409347

[82] Cho YI, Choi BG. Validation of an electronic anti-fouling technology in a single-tube heat exchanger. International Journal of Heat and Mass Transfer. 1999;**42**:1491-1499. DOI: 10.1016/S0017-9310(98)00196-3

[83] Grandison C, Piola R, Fletcher L. A Review of Marine Growth Protection System (MGPS) Options for the Royal Australian Navy. DSTO-TR-2631. 2011;**38**:2011. Available from: http://www.dtic.mil/dtic/tr/fulltext/u2/a557490.pdf

[84] Pouponnot F. Heat exchanger tube inserts an update in new applications with trouble shooting aspects in crude units, residue service, reboilers, u-tubes. In: Proceedings of the 6th International Conference on Heat Exchanger Fouling and Cleaning-Challenges and Opportunities. Engineering Conferences International. 2005. ISBN: 978-1-5108-1946-7

[85] Xiaokai X. Research on the electromagnetic anti-fouling technology for heat transfer enhancement. Applied Thermal Engineering. 2008;**28**:889-894. DOI: 10.1016/j.applthermaleng.2007.07.006

[86] Andritsos N, Karabelas AJ. Calcium carbonate scaling in a plate heat exchanger in the presence of particles. International Journal of Heat and Mass Transfer. 2003;**46**:4613-4627. DOI: 10.1016/S0017-9310(03)00308-9

[87] Lipus LC, Krope J, Crepinsek L. Dispersion destabilization in magnetic water treatment. Journal of Colloid and Interface Science. 2001;**236**:60-66. DOI: 10.1006/jcis.2000.7392

[88] Gabrielli C, Jaouhari R, Maurin G, Keddam M. Magnetic water treatment for scale prevention. Water Research. 2001;**35**:3249-3259. DOI: 10.1016/S0043-1354(01)00010-0

[89] Wilson DI. Heat exchanger fouling and cleaning: Fundamentals and applications. Challenges in Cleaning: Recent Developments and Future Prospects, Engineering Conferences International. ECI Digital Archives. 2003

Chapter 5

Molecular Dynamics Simulation-Based Study on Enhancing Thermal Properties of Graphene-Reinforced Thermoplastic Polyurethane Nanocomposite for Heat Exchanger Materials

Animesh Talapatra and Debasis Datta

Abstract

Molecular dynamics (MD) simulation-based development of heat resistance nanocomposite materials for nanoheat transfer devices (like nanoheat exchanger) and applications have been studied. In this study, MD software (Materials Studio) has been used to know the heat transport behaviors of the graphene-reinforced thermoplastic polyurethane (Gr/TPU) nanocomposite. The effect of graphene weight percentage (wt%) on thermal properties (e.g., glass transition temperature, coefficient of thermal expansion, heat capacity, thermal conductivity, and interface thermal conductance) of Gr/TPU nanocomposites has been studied. Condensed-phase optimized molecular potentials for atomistic simulation studies (COMPASS) force field which is incorporated in both amorphous and forcite plus atomistic simulation modules within the software are used for this present study. Layer models have been developed to characterize thermal properties of the Gr/TPU nanocomposites. It is seen from the simulation results that glass transition temperature (Tg) of the Gr/TPU nanocomposites is higher than that of pure TPU. MD simulation results indicate that addition of graphene into TPU matrix enhances thermal conductivity. The present study provides effective guidance and understanding of the thermal mechanism of graphene/TPU nanocomposites for improving their thermal properties. Finally, the revealed enhanced thermal properties of nanocomposites, the interfacial interaction energy, and the free volume of polymer nanocomposites are examined and discussed.

Keywords: molecular dynamics (MD) simulations, thermo-mechanical properties, glass transition temperature, coefficient of thermal expansion, thermal conductivity

1. Introduction

Optimum design of heat exchanger using nanotechnology is a burning field to reduce the energy consumption. Recently, application of nanosolids, nanofluids, and nanogases is the promising nanolevel research area of interest for energy savings in heat exchanger. Investigation of the nanolevel heat transfer using molecular dynamics (MD) simulation is only a new pioneer concept for the last few years [1–5]. Here investigations are based on atomic movement within a nanosystem during MD simulation. Experimental study on the heat transfer of nano-size devices or equipments are very time-consuming and expensive for the existing testing capabilities [6–9]. Many studies have been done for enhancing the energy consumption of heat exchanger by improving thermal properties. This thermal behavior of nanocomposite can bring a huge transformation and innovation in the heat transfer. The ultrathin thermal polyurethane heat transfer material can be applied to a number of different fabrics in heat exchanger. The use of graphene (Gr) in thermoplastic polyurethane (TPU), that is, Gr/TPU nanocomposite in place of traditional material in heat exchanger, increases the heat transfer rate in a significant manner. Hussein studied to calculate thermal properties of metals and nonmetals at room temperature for applications in heat exchanger and found that metallic materials are preferably suitable for heat transfer application [10]. But there is some limitation for application of metals due to corrosion. To overcome the situation, implementation of polymer heat exchanger technology for the past decades is a pioneering innovation for heat exchanger materials [11]. The major limitation of polymer for application in heat exchanger is very low thermal conductivity. To improve that property graphene-reinforced polymer nanocomposite is a suitable candidate material in evaporation and condensation applications within heat exchanger [12]. Thermal performance of polymer nanocomposite heat exchangers mainly deals with on shell and tube heat exchangers, plate heat exchangers, finned tube heat exchangers, immersed tube heat exchangers, and hollow fiber heat exchangers [13]. Currently, thermoplastic elastomer is used in heat exchanger applications. Thermoplastics elastomer can be repeatedly softened by heating and solidified by cooling as long as the material is not thermally damaged by overheating [14]. The thermal expansion of thermoplastic polymer can be beneficial with regard to fouling because repeated expansion and contraction of the polymer channels can lead to scale detachment [15]. It is seen from previous studies that there are less number of simulation-based studies like MD simulation on enhanced heat transfer in heat exchanger materials. Detail results of MD simulation are obtained by solving Newton's equation of motion of every atom within nanoscopic system. The basic dynamics parameters of all atoms, that is, position, velocity, and interaction force, play a vital rule during MD simulation. Nanoheat transfer problems of nanocomposites are related to thermo-mechanical properties of nanomaterials. For the design of nanodevices like nanoheat exchanger (NHE), the concepts of the nanothermal properties with temperature variation and dimension of the nanodevice is very much promising ideas. Determination of the thermal properties of nanocomposites by MD simulations is very time-consuming and a challenging task. The objective of the chapter is to characterize thermal properties like thermal conductivity and thermal expansion coefficient of graphene-reinforced polyurethane nanocomposite using molecular dynamics (MD) modeling for nanoheat exchanger material application.

MD simulation consists of many parts which are mainly (a) molecular interactions, (b) molecular minimization, (c) algorithms, (d) ensemble, (e) boundary conditions, (f) atomistic stress calculation, etc. MD simulation helps to determine the position (r_i) and velocity (v_i) vectors of atom i with the time integration method (e.g., the velocity Verlet algorithm):

$$r_i(t + \Delta t) = r_i(t) + v_i(t)\Delta t + \frac{1}{2}a_i(t)\Delta t^2 \quad (1)$$

$$v_i(t+\Delta t) = v_i(t) + \frac{a_i(t) + a_i(t + t\Delta t)}{2}\Delta t \tag{2}$$

where $(\boldsymbol{a}_i)$ is the acceleration of atom i, t is the current time, and Δt is the time step.

$$a_i = \frac{F_i}{m_i} \tag{3}$$

where m_i is mass of atom ith and $\boldsymbol{F}_i$ is the force vector of atom ith obtain from the gradient of the total potential energy (E) on atom i

$$F_i = \frac{dE}{dr_i} \tag{4}$$

In MD simulation, a model system is built at the atomistic level with prescribed potentials (also known as the force field) acting between the atoms. The potential energy is dependent on the force field that is applied to the system. The potential energies of the system are determined from both bonded and nonbonded energies. The total potential energy combines all energetic contributions shown in the following equation:

$$\mathrm{E} = \mathrm{E_{bonded}} + \mathrm{E_{non\text{-}bonded}} \tag{5}$$

$$\mathrm{E} = (\mathrm{E_{valance}} + \mathrm{E_{cross\text{-}term}}) + \mathrm{E_{non\text{-}bonded}} \tag{6}$$

$$\mathrm{E} = (\mathrm{E_{bond}} + \mathrm{E_{angle}} + \mathrm{E_{torsion}} + \mathrm{E_{inversion}}) + \mathrm{E_{cross\text{-}term}} + (\mathrm{E_{vdW}} + \mathrm{E_{coulomb}}) \tag{7}$$

The different types of bond and nonbond energy equations have been shown below:

$$E_{bond} = \frac{1}{2}k_b(b - b_0)^2 \tag{8}$$

where k_b is the stretching force constant, b_0 is the equilibrium bond length, and b is the actual bond length.

$$E_{bend} = \frac{1}{2}k_\theta(\theta - \theta_0)^2 \tag{9}$$

where k_θ is the angle-bending force constant, θ_0 is the equilibrium bond angle, and θ is the actual bond angle.

$$E_{torsion} = \frac{1}{2}k_\phi(1 + \cos(n\phi - \phi_0)) \tag{10}$$

where $k\varphi$ is the torsional barrier, φ is the actual torsion angle, n is the periodicity, and φ_0 is the reference torsional angle.

$$E_{inversion} = \frac{1}{2}k_\omega(\omega - \omega_0)^2 \tag{11}$$

where K_ω is the force constant and ω is the angle between the axis and the plane.

$$E_{vdw} = \sum \frac{A_{ij}}{r_{ij}^{12}} - \frac{B_{ij}}{r_{ij}^{6}} \tag{12}$$

where A_{ij} and B_{ij} are the repulsive and attractive term coefficients, respectively, and r_{ij} is the distance between the two atoms.

$$E_{coulomb} = \frac{1}{\varepsilon}\frac{q_1 q_2}{r_{ij}} \tag{13}$$

where q_1 and q_2 are the charges on the interacting atoms, ε is the dielectric constant, and r_{ij} is the interatomic distance.

The Condensed-phase Optimized Molecular Potentials for Atomistic Simulation Studies (COMPASS) [16] which is incorporated in both amorphous and forcite plus atomistic simulation modules in the Material Studio is used for this present study. COMPASS functional form has 11 valence terms (including diagonal and off-diagonal cross coupling terms) and 2 nonbond interaction terms (the Coulombic and Lennard-Jones functions for electrostatic and van der Waals (vdW) interactions, respectively). During all the simulations, the temperature and pressure are maintained by Andersen and Berendsen method. The calculation of nonbonded interactions is simulated by applying a cutoff distance of 12.5 Å. The spline and buffer widths are 1 and 0.5 Å, respectively.

Experimental methods for prediction of the enhanced thermal properties of graphene-reinforced nanocomposites are limited because nanometer scale measurements are difficult and costly. Thus, MD simulation techniques are only an economical path to characterize nanomaterial and nanocomposites for heat exchanger material within small length and small time.

2. MD simulation models and methods for thermal property calculation

Heat can transfer through electrons and phonons, by excitations and by scattering in the nanocomposites [17]. This different ways heat transfer methods help to understand the mechanism of thermal properties enhancement in graphene-based nanocomposites. In order to study the enhanced thermal properties of graphene-reinforced thermoplastic polyurethane nanocomposites, the wide range thermal parameters of the nanocomposite material (like thermal conductivity, coefficient of the thermal expansion, glass transition temperature, etc.) have to be calculated. To calculate these parameters, different simulation models are constructed using molecular modeling software package by Materials Studio 2017. In MD simulation study, there are three types of models which are developed, namely concentrated model (CM), layer model (LM), and interfacial model (IM). Different models have been used to study different properties of the nanocomposite with respect to different parameters. In this study, we have focused mainly on layer models to characterize enhanced thermal properties. Both graphene and polyurethane models are simulated separately before constructing the graphene-reinforced thermoplastic polyurethane (Gr/TPU) nanocomposites. The condensed-phase optimized molecular potential for atomistic simulation studies (COMPASS) force field has been selected to describe the atomistic behavior for the simulation models. In MD simulation, each atom is modeled as a point mass and interacts with other atoms through force field. The position and momentum of atoms are updated based on Newton's equation of motion.

After the construction all of the graphene and TPU model using amorphous cell module within the Material Studio, build layer option is used to construct the layer models of the Gr/TPU nanocomposites. It is seen that with high weight fraction condition of graphene, nanocomposites behave likely more brittle than polymer matrix due to growth of void and chain disentanglement. So, 1% weight fraction of

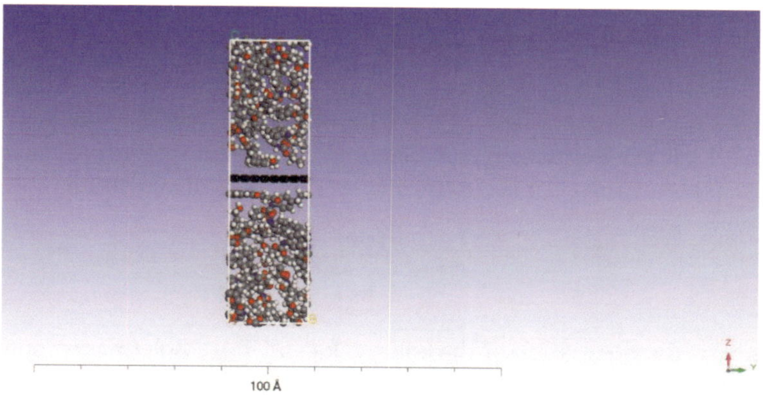

Figure 1.
Developed 1% Gr/TPU nanocomposite model.

graphene-reinforced nanocomposites is considered in this study. The constructed nanocomposite lattice parameters with dimensions a = 22 Å, b = 22 Å, and c = 95 Å are constrained in such a way that after dynamic equilibrium process density of Gr/TPU, nanocomposite is 1.38 g/cc (nearly experimental value). MD simulation run is divided into two parts, namely equilibration run and production run. Equilibration run helps to develop the molecular structure under the condition of the desired thermodynamic state, while the production run helps to calculate different thermal parameters, namely specific heat, thermal expansion, and thermal conductivity. In equilibration run, two important conditions have to be fulfilled. One is the minimum energy stabilized condition at a prescribed temperature, and another is initial stress-free structure within periodic boundary condition. So, first steps nanocomposite models are moved through energy minimization, canonical ensemble (constant number of atoms, volume, and temperature) (NVT)) dynamic simulations, and temperature annealing cycle, respectively. The duration of the dynamic run is considered 200 ps with an integration time step of 1 fs (femtosecond). This process is followed by graphene as a rigid structure so that the lattice dimension (c) in the z-direction will be changed. Temperature annealing cycle involves temperature up and down from 300 to 600 K to get the minimization of energy in the structure. The annealing time is set for 500 ps, during which the temperature is raised from 300 to 600 K with a rate of 6 K/ps and cool down to 300 K with the same rate. In the second step, the non-constrained parts (TPU) within lattices are compressed in such a way so that the final density of nanocomposite will be 1.38 g/cc (nearly experimental value) after using isothermal-isobaric (NPT) ensembles. The lowest energy structure models are fully relaxed under an isothermal-isobaric NPT ensemble (i.e., constant numbers of atoms, pressure, and temperature) at 300 K and 1 atm for 500 ps. The isothermal-isobaric (NPT) ensembles help to relax the lattices parameters and angles in order to obtain a final reasonable equilibrated structure. These steps generate various curves of various parameters such as energy, pressure, volume, and temperature versus simulation run time. These curves are very important to study thermal properties of the nanocomposite. **Figure 1** shows the developed 1% Gr/TPU nanocomposite model by MD simulation.

After dynamic equilibrium process density of 1% Gr/TPU, the nanocomposite is 1.38 g/cc which is shown in **Figure 2**.

Heat capacity is one of the important thermal properties for the nanocomposite system. In this work, MD simulation is applied to calculate the isobaric heat capacity (C_P), and the value of Cp can be determined according to the following equation:

$$Cp = \frac{(KE + PE + PV)^2}{K_B\,T} \tag{14}$$

where KE is the kinetic energy, PE is the potential energy, P is the pressure, V is the volume, K_B is the Boltzmann constant, and T is the temperature. The specific heat at constant volume (C_v) is obtained from the following equation:

$$Cv = \frac{\left(\delta E^2\right)_{NVT}}{k_B T^2} \tag{15}$$

where δE is the fluctuation of the energy, k_B and T are Boltzmann constant, and absolute temperature, respectively.

In order to study the glass transition temperature and coefficient of thermal expansion (CTE), a high-temperature annealing protocol is followed. At each temperature, the system is equilibrated by isothermal-isobaric (NPT) ensemble in MD simulation at atmospheric pressure for 500 ps. The temperature is raised up to 600 K and equilibrated for 500 ps using an NPT ensemble under atmospheric pressure and then dropped by 20 K each time until it reached 300 K. The cooling down method is applied after the heating up method by decreasing the temperature with the same settings and simulation time. Since each temperature drop is only 20 K, the structure is re-equilibrated very quickly every time its temperature is decreased. For each temperature, the volume of the simulation box V is examined over the duration time of the MD simulation, and the average value is calculated. From the volume versus temperature relationship curve, it is seen that there is a discontinuity in the volume versus temperature slope, which gives the glass transition temperature (Tg) of nanocomposite. The volume versus temperature (V-T) results are important to know two factors; first, this provides a means of determining the quality of the force field used in the simulations, and second, prediction of the volumetric glass transition temperature (Tg) [18]. The simulation result is in good agreement with experiment and demonstrates the accuracy of COMPASS force field. The volumetric coefficient of thermal expansion (VCTE) is defined by (α) [19]:

$$\alpha = \frac{1}{V_0}\frac{\Delta V}{\Delta T} \tag{16}$$

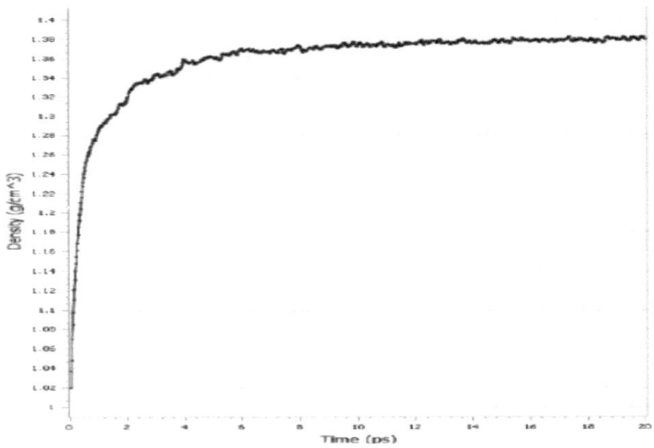

Figure 2.
Density (g/cc) versus time (ps) in dynamic equilibrium run.

where V_0 is the equilibrated system volume before the cooling simulation starts. The fractional change of volume with respect to temperature ($\Delta V/\Delta T$) is obtainable from volume versus temperature relationship curve. It is seen that change of volume with respect to temperature ($\Delta V/\Delta T$) is a different value above glass transition temperature (Tg) for graphene-reinforced thermoplastic polyurethane nanocomposite. So, volumetric coefficient of thermal expansion (VCTE) has two values due to glass transition temperature (Tg) [20]. The glass transition temperature and volumetric coefficient of thermal expansion (VCTE) of nanocomposite can also be obtained by calculating the free volume as a function of temperature, since the free volume undergoes an abrupt change when the material goes through glass transition. By probing the lattice cell with a spherical probe, using the "atom volume and surfaces" tool of the Materials Studio (MS), the free volume in the nanocomposite is calculated as a function of temperature during the annealing process as shown in **Figure 3**.

Free volume is the volume that is not occupied by either the graphene or the TPU chains. The free volume fraction (FVF) can be obtained by the following equation [21]:

$$FVF = \frac{V_f}{V_f + V_o} \tag{17}$$

where V_f is the free volume and V_0 is the occupied volume of the polymer chains.

Thermal conductivity is the sum of the phonon contribution and the electronic contribution. Therefore, total thermal conductivity (K)

$$K = K_e + K_p \tag{18}$$

where K_p and K_e are the phonon contribution and the electronic contribution to the thermal conductivity, respectively. Though, electrons contributed thermal conductivity is neglected in most graphene-reinforced nanocomposite. Thermal conductivity is an important thermal property relevant to thermal management applications. Thermal conductivity is generally calculated using equilibrium or nonequilibrium MD approaches. Equilibrium molecular dynamic (EMD) facilitates thermal conductivity prediction in all directions using one simulation, whereas nonequilibrium molecular dynamic (NEMD) requires the use of a thermal gradient and therefore only enables the calculation of thermal conductivity in one

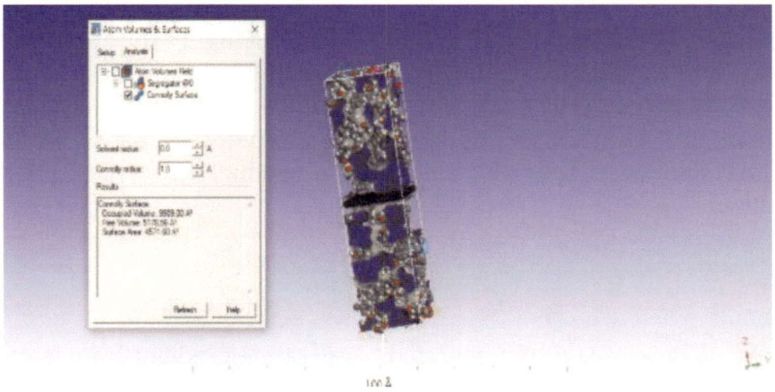

Figure 3.
Free volume calculation using atom volume and surfaces tool in MS.

direction. The indirect method is an equilibrium molecular dynamic (EMD) method which is derived from Green-Kubo approach [22, 23], where current fluctuations are used to compute the thermal conductivity via the fluctuation-dissipation theorem [24]:

$$K_{\alpha\beta} = \frac{V}{k_B T^2}\int_0^\infty \langle J_\alpha(t) J_\beta(0)\rangle dt \tag{19}$$

where k_B is the Boltzmann constant, V and T denote the volume and temperature of the system, J_α is the heat flux in the α direction, and the angular brackets denote the ensemble average. The heat flux vector can be written as

$$\mathrm{J(t)} = \frac{1}{V}\frac{d}{dt}\sum_{i=1}^{N} r_i E_i \tag{20}$$

where r_i and E_i are the position and total energy of the ith atom, respectively. EMD is particularly useful for geometries where periodic boundary conditions can be applied. EMD is often computationally more expensive, and the results are more sensitive to the simulation parameters. In EMD, the system is set to the desired temperature, and then a constant energy scheme is used with the well-known Green-Kubo relations to calculate the thermal conductivity tensor. The direct method is a NEMD method in which a temperature difference is introduced into the simulation domain and the thermal conductivity is computed according to Fourier's law as $K = -J/\Delta T$, where J and ΔT are heat flux and temperature gradient across the system, respectively. For nonequilibrium MD (NEMD) methods, a long slab of polymer nanocomposite is constructed, and a difference in temperature is established between a heat source and a sink at the ends of the slab, and the flux is calculated. Equilibrium systems are simulated by nonequilibrium MD (NEMD) based on our homemade PERL script to compute their thermal conductivities. The nonequilibrium state can be established either by applying two thermostats at different temperatures to maintain a constant temperature at the two ends of the system or by artificially swapping atom velocities in different regions to impose a constant heat flux also known as the reverse nonequilibrium MD (RNEMD) method based on Muller-Plathe's approach [25]. In the reverse nonequilibrium MD (RNEMD) method, the energy exchange is carried out by exchanging the kinetic energy of two particles: the hottest particle in the cold layer and the coldest particle in the hot layer. The energy E is therefore variable and needs averaging over many exchanges. In the RNEMD method, the simulation box was divided into a number of slabs in the heat flux (z) direction with the same thickness. The heat flux was generated by exchanging the kinetic energy between the highest kinetic (the hottest) atom in the heat sink and the lowest kinetic energy (the coldest) atom in the heat source. The larger momentum exchange rate in RNEMD method suggests higher energy exchange frequency between heat source and heat sink. The thermal conductivity was calculated using Fourier's law:

$$K = -\frac{J_q}{\nabla T} \tag{21}$$

where ∇T denoted the time-integrated temperature gradient from least squares approximation of the discrete temperatures according to the heat flow direction. The temperature of each slab was calculated using the virial theorem, and J_q is the heat flux given as

$$J_q = \frac{E}{2A_c\Delta t} \quad (22)$$

where E is the subtracted energy from the heat sink. A_c is cross-sectional area and Δt are the time step.

Interface thermal conductance (ITC) is developed between graphene and polymer matrix within nanocomposite due to their weak interactions. However, the thermal conductivity (TC) of graphene-reinforced TPU nanocomposites are far below the thermal conductivity (TC) of graphene. Such a low filler efficiency is most likely attributed to the low interface thermal conductance (ITC) between graphene and polymer. The simulating method (pump-probe transient thermoreflectance method) is used to calculate the interface thermal conductance (ITC) between the filler (graphene) and matrix (TPU). The graphene-TPU interfacial thermal resistance (R) is then calculated by following equation:

$$R = \frac{\Delta T}{J_q} \quad (23)$$

3. Results and discussion

In the equilibration stage within MD simulation, NPT dynamic run is carried out for 500 ps at room temperature and pressure to generate curves of energy, density, pressure, and temperature versus time. These curves are used to decide the cutoff between equilibration and production runs. It is observed that all the models are equilibrated at 50 ps, that is, no fluctuations after 50 ps. At the end of the equilibrations, the density of the nanocomposites stabilized at an average density of 1.3 g/cc with a standard deviation of 0.02 g/cc. The reason behind for the density differs from experimental value because MD simulation deals with material defect-free and impurities. It is seen in the volume versus temperature (V-T) curve during NPT dynamic run that free volume change affects the thermal properties of the nanocomposites. Further it is also observed that free volume within lattice increases according to the strain application. The glass transition mainly depends on two factors: (a) free volume and (b) the mobility of chain segments. Initially there is no graphene in TPU chains for that free volume is zero in the simulated cell and polymers are free to move within this cell volume. As a result, high values for the entropy and low values for the bulk, shear, and Young's moduli. Further, the incorporation of the graphene into the simulated cells increases free volume and decrease the entropy and increase the values of the bulk, shear, and Young's moduli. Since the entropy of the system is related to the free volume and the Connolly surface of the TPU nanocomposites, the prediction of these parameters may give a concept on the enhancement of the nanocomposite thermal properties. The simulation results show that the TPU matrix reinforced with graphene have a tendency to increase glass transition temperature (Tg) for stronger interlocking between the graphene and TPU molecules. The Connolly surface to volume ratio is small values for the neat TPU and increase with the increase in the graphene loading. The free volume is defined as the volume on the side of the Connolly surface without atoms. This simulation study reveals that Tg of TPU is in the range of 285–305°C and for the Gr/TPU nanocomposite is 350 K (experimental 223–323 K). From the V-T curve, volumetric coefficient of thermal expansion (α) is evaluated from the slope above and below Tg, which is between 2.6×10^{-5} and $2.4 \times 10^{-4}/K^{-1}$ (experimental $3.15 \times 10^{-4}/K^{-1}$). The free volume change rapidly when the material goes through glass transition, according to Fox and Flory's theory of glass transition [20].

Further simulation study reveals that graphene/TPU nanocomposite thermal conductivity is 1.5 W/mK, whereas TPU thermal conductivity is 0.2 W/mK. NEMD simulations are used to calculate the thermal conductivity either by imposing a thermal gradient into the system of particles or by introducing a heat flux flow in the reverse nonequilibrium MD (RNEMD) method. In the present study, heat flux flow method is used to calculate thermal conductivity in the longitudinal direction. The total number of layers is 40. Two types of exchange method are used in the present study, namely VARIABLE and FIXED. About 1 kcal/mol energy exchange is taken in FIXED method. The number of exchanges is taken as 500 for equilibrium stage under NVT and 1000 for production stage under NVE. The time steps in between two exchanges are fixed at 100. Due to the presence of the graphene-TPU interface, there exists a temperature jump ΔT at the interface. The present study obtained values are in good agreement with previous values obtained from simulations and experimental measurements in the literatures [26–31]. The present study contributes some novel procedures during MD simulation work which will not be done in previous researchers.

4. Conclusions

After all earlier studies, it can be concluded that the development of new technologies are giving a new attention on to investigate nanoscale phenomena (including nanoscale heat transfer). Therefore, MD simulation is the only nanoscale tool to investigate the enhancement of thermal properties of graphene-reinforced nanocomposites for heat exchanger material. Based on the current simulation results, it is found that graphene-reinforced TPU nanocomposites demonstrate higher moduli, higher glass transition temperature, and lower values of CTE than pure TPU, that is, without reinforcements. This provides useful information to understand the nanoheat transport behaviors within TPU nanocomposites for the future development of thermal nanodevice. By taking advantage of low-cost simulations to establish material designs, overall materials development costs can be dramatically reduced, and development times can be expedited.

Acknowledgements

The authors would like to thank the organizer committee members of TEQIP which funded short-term course on mechanics of composite using Material Studio in NIT, Durgapur, and further license software support to conduct MD simulation. Thanks to nanoHUB Pro for instructions and online helps. Also thanks to Accelrys (recently BIOVIA) Materials Studio community members help for simulation study.

Nomenclature

E	total potential energy
k_b	the stretching force constant
k_θ	the angle-bending force constant
k_φ	the torsional barrier
K_ω	the force constant
T	thermodynamic temperature
V	volume of nanocomposites

C_p	specific heat of nanoparticle bulk material
k	thermal conductivity
k_B	Boltzmann constant
K_p	the phonon contribution
K_e	the electronic contribution
J_q	the heat flux
R	interfacial thermal resistance
α	the volumetric coefficient of thermal expansion
ρ	density of nanomaterial
r_i	the position of the ith atom
E_i	total energy of the ith atom

Author details

Animesh Talapatra[1*] and Debasis Datta[2]

1 MCKVIE, Howrah, West Bengal, India

2 IIEST, Howrah, West Bengal, India

*Address all correspondence to: animesh_talapatra@yahoo.co.in

References

[1] Komarneni S. Feature article. Nanocomposites. Journal of Materials Chemistry. 1992;**2**:1219-1230

[2] Jordan J, Jacob KI, Tannenbaum R, Sharaf MA, Jasiuk IM. Experimental trends in polymer nanocomposites—A review. Materials Science and Engineering A. 2005;**393**(1–2):1-11. DOI: 10.1016/j.msea.2004.09.044

[3] Podsiadlo P, Tang Z, Shim BS, Kotov NA. Counterintuitive effect of molecular strength and role of molecular rigidity on mechanical properties of layer-by-layer assembled nanocomposites. Nano Letters. 2007; **7**(5):1224-1231

[4] Chan L, Zuo X, Wang L, Wang E, Song S, Wang J, et al. Flexible carbon nanotube-polymer composite films with high conductivity and super hydrophobicity made by solution process. Nano Letters. 2008;**8**(12): 4454-4458

[5] Lehn JM, Fendler JH, Meldrum F. The colloid chemical approach to nanostructured materials. Advanced Materials. 1995;**7**(7):607-632

[6] Seol JH, Jo I, Moore AL, Lindsay L, Aitken ZH, Pettes MT, et al. Two-dimensional phonon transport in supported graphene. Science. 2010; **328**(5975):213-216

[7] Klemens PG, Pedraza DF. Thermal conductivity of graphite in the basal plane. Carbon. 1994;**32**(4):735-741

[8] Allen PB, Feldman JL. Thermal conductivity of glasses: Theory and application to amorphous Si. Physical Review Letters. 1989;**62**(6):645

[9] Allen PB, Feldman JL, Fabian J, Wooten F. Diffusons, locons and propagons: Character of atomic vibrations' in amorphous Si. Philosophical Magazine B. 1999;**79** (11–12):1715-1731

[10] Hussein AM, Bakar RA, Kadirgama K, Sharma KV. Experimental measurement of nanofluids thermal properties. International Journal of Automotive and Mechanical Engineering. 2013;**7**:850-863

[11] Cevallos JG, Bergles AE, Bar-Cohen A, Rodgers P, Gupta SK. Polymer heat exchangers-history, opportunities, and challenges. Heat Transfer Engineering. 2012;**33**(13):1075-1093

[12] Chen H, Ginzburg VV, Yang J, Yang Y, Liu W, Huang Y, et al. Thermal conductivity of polymer-based composites: Fundamentals and applications. Progress of Polymer Science. 2016;**59**:41-85

[13] Chen X, Su Y, Reay D, Riffat S. Recent research in polymer heat exchangers—A review. Renewable and Sustainable Energy Reviews. 2016;**60**: 1367-1386

[14] T'Joen C, Park Y, Wang Q, Sommers A, Han X, Jacobi A. A review on polymer heat exchangers for HVAC&R applications. International Journal of Refrigeration. 2009;**32**: 763-779

[15] Zarkadas DM, Sirkar KK. Polymeric hollow fiber heat exchangers: An alternative for lower temperature applications. Industrial and Engineering Chemistry Research. 2004;**43**: 8093-8106

[16] Sun H. COMPASS: An ab initio force-field optimized for condensed-phase applications overview with details on alkane and benzene compounds. The Journal of Physical Chemistry B. 1998; **102**(38):7338-7364

[17] Xu Z. Heat transport in low-dimensional materials: A review and perspective. Theoretical and Applied Mechanics Letters. 2016;**6**(3):113-121

[18] Choi J, Yu S, Yang S, Cho M. The glass transition and thermo elastic behavior of epoxy-based nanocomposites: A molecular dynamics study. Polymer. 2011;**52**(22):5197-5203

[19] Shiu SC, Tsai JL. Characterizing thermal and mechanical properties of graphene/epoxy nanocomposites. Composites Part B: Engineering. 2014; **56**:691-697

[20] Yang S, Qu J. Computing thermo mechanical properties of cross-linked epoxy by molecular dynamic simulations. Polymer. 2012;**53**(21): 4806-4817

[21] Schmidtke E, Günther-Schade K, Hofmann D, Faupel F. The distribution of the unoccupied volume in glassy polymers. Journal of Molecular Graphics & Modelling. 2004;**22**(4):309-316

[22] Green MS. Mark off random processes and the statistical mechanics of time-dependent phenomena. II. Irreversible processes in fluids. The Journal of Chemical Physics. 1954;**22**(3): 398-413

[23] Kubo R. Statistical-mechanical theory of irreversible processes. I. General theory and simple applications to magnetic and conduction problems. Journal of the Physical Society of Japan. 1957;**12**(6):570-586

[24] Sellan DP, Landry ES, Turney JE, Mc Gaughey AJH, Amon CH. Size effects in molecular dynamics thermal conductivity predictions. Physical Review B. 2010;**81**:214-305

[25] Müller-Plathe F. A simple nonequilibrium molecular dynamics method for calculating the thermal conductivity. The Journal of chemical physics. 1997;**106**(14):6082-6085

[26] Fox TG, Loshaek S. Influence of molecular weight and degree of crosslinking on the specific volume and glass temperature of polymers. Journal of Polymer Science. 1955;**15**(80):371-390

[27] Wu SL, Shi TJ, Zhang LY. Latex co-coagulation approach to fabrication of polyurethane/graphene nanocomposites with improved electrical conductivity, thermal conductivity, and barrier property. Journal of Applied Polymer Science. 2016;**133**(11):13

[28] Lee SH, Jung JH, Oh IK. 3D networked graphene-ferromagnetic hybrids for fast shape memory polymers with enhanced mechanical stiffness and thermal conductivity. Small. 2014; **10**(19):3880-3886

[29] Li A, Zhang C, Zhang YF. Thermal conductivities of PU composites with graphene aerogels reduced by different methods. Composites Part A: Applied Science and Manufacturing. 2017;**103**: 161-167

[30] Yadav SK, Cho JW. Functionalized graphene nanoplatelets for enhanced mechanical and thermal properties of polyurethane nanocomposites. Applied Surface Science. 2013;**266**:360-367

[31] Materials Studio. User's Manual, Version 1.2. San Diego, CA: Accelrys, Inc.; 2001

Section 2

Inverse Problem

Chapter 6

A Numerical Approach to Solving an Inverse Heat Conduction Problem Using the Levenberg-Marquardt Algorithm

Tao Min, Xing Chen, Yao Sun and Qiang Huang

Abstract

This chapter is intended to provide a numerical algorithm involving the combined use of the Levenberg-Marquardt algorithm and the Galerkin finite element method for estimating the diffusion coefficient in an inverse heat conduction problem (IHCP). In the present study, the functional form of the diffusion coefficient is an unknown priori. The unknown diffusion coefficient is approximated by the polynomial form and the present numerical algorithm is employed to find the solution. Numerical experiments are presented to show the efficiency of the proposed method.

Keywords: parabolic equation, inverse problem, Levenberg-Marquardt

1. Introduction

The numerical solution of the inverse heat conduction problem (IHCP) requires to determine diffusion coefficient from an additional information. Inverse heat conduction problems have many applications in various branches of science and engineering, mechanical and chemical engineers, mathematicians and specialists in many other sciences branches are interested in inverse problems, each with different application in mind [1–15].

In this work, we propose an algorithm for numerical solving an inverse heat conduction problem. The algorithm is based on the Galerkin finite element method and Levenberg-Marquardt algorithm [16–17] in conjunction with the least-squares scheme. It is assumed that no prior information is available on the functional form of the unknown diffusion coefficient in the present study, thus, it is classified as the function estimation in inverse calculation. Run the numerical algorithm to solve the unknown diffusion coefficient which is approximated by the polynomial form. The Levenberg-Marquardt optimization is adopted to modify the estimated values.

The plan of this paper is as follows: in Section 2, we formulate a one-dimensional IHCP. In Section 3, the numerical algorithm is derived. Calculation of sensitivity coefficients will be discussed in Section 4. In order to discuss on some numerical aspects, two examples are given in Section 5. Section 6 ends this paper with a brief discussion on some numerical aspects.

2. Description of the problem

The mathematical formulation of a one-dimensional heat conduction problem is given as follows:

$$\frac{\partial u}{\partial t} = \frac{\partial}{\partial x}\left[q(x)\frac{\partial u}{\partial x}\right] + f(x,t), \quad (x,t) \in (0,\mathrm{L}) \times (0,\mathrm{T}], \tag{1}$$

with the initial condition

$$u(\mathrm{x},0) = \mathrm{u}_0(x), \quad 0 \le x \le L, \tag{2}$$

and Dirichlet boundary conditions

$$u(0,\mathrm{t}) = \mathrm{g}_1(t), \quad 0 \le t \le T, \tag{3}$$

$$u(1,\mathrm{t}) = \mathrm{g}_2(t), \quad 0 \le t \le T, \tag{4}$$

where $\mathrm{f}(\mathrm{x},\mathrm{t}), u_0(\mathrm{x}), g_1(\mathrm{t}), g_2(\mathrm{t})$ and $q(\mathrm{x})$ are continuous known functions. We consider the problem (1)–(4) as a direct problem. As we all know, if $u_0(\mathrm{x}), g_1(\mathrm{t})$, $g_2(\mathrm{t})$ are continuous functions and $q(\mathrm{x})$ is known, the problem (1)–(4) has a unique solution.

For the inverse problem, the diffusion coefficient $q(\mathrm{x})$ is regarded as being unknown. In addition, an overspecified condition is also considered available. To estimate the unknown coefficient $q(\mathrm{x})$, the additional information on the boundary $x = x_0$, $0 < x_0 < L$ is required. Let the $u(\mathrm{x},\mathrm{t})$ taken at $x = x_0$ over the time period $[0,\mathrm{T}]$ be denoted by

$$\mathrm{u}(x_0,t) = \mathrm{g}(t) \quad 0 \le t \le T. \tag{5}$$

It is evident that for an unknown function $q(\mathrm{x})$, the problem (1)–(4) is under-determined and we are forced to impose additional information (5) to provide a unique solution pair $(\mathrm{u}(\mathrm{x},\mathrm{t}), \mathrm{q}(\mathrm{x}))$ to the inverse problem (1)–(5).

We note that the measured overspecified condition $\mathrm{u}(\mathrm{x}_0,\mathrm{t}) = \mathrm{g}(t)$ should contain measurement errors. Therefore the inverse problem can be stated as follows: by utilizing the above-mentioned measured data, estimate the unknown function $q(\mathrm{x})$.

In this work the polynomial form is proposed for the unknown function $q(\mathrm{x})$ before performing the inverse calculation. Therefore $q(\mathrm{x})$ approximated as

$$q(x) \approx \hat{q(\mathrm{x})} = \mathrm{p}_1 + p_2 x + p_3 x^2 + \cdots + p_{m+1} x^m, \tag{6}$$

where $p_1, p_2, \cdots, p_{m+1}$ are constants which remain to be determined simultaneously. The unknown coefficients $p_1, p_2, \cdots, p_{m+1}$ can be determined by using least squares method. The error in the estimate

$$F(p_1, p_2, \ldots, p_{m+1}) = \sum_{i=1}^{n} \left[u(x_0, t_i, p_1, p_2, \ldots, p_{m+1}) - g(\mathrm{t}_i)\right]^2, \tag{7}$$

is to be minimized. Here, $u(x_0, t_i, p_1, p_2, \ldots, p_{m+1})$ are the calculated results. These quantities are determined from the solution of the direct problem which is given previously by using an approximated $\hat{q(\mathrm{x})}$ for the exact $q(\mathrm{x})$. The estimated values of p_j, $j = 1, 2, \cdots, m+1$ are determined until the value of $F(p_1, p_2, \ldots, p_{m+1})$ is minimum. Such a norm can be written as

$$F(\mathbf{P}) = [\mathbf{U}(\mathbf{P})\text{-}\mathbf{G}]^T[\mathbf{U}(\mathbf{P})\text{-}\mathbf{G}], \tag{8}$$

where $\mathbf{P}^T = [p_1, p_2, \cdots, p_{m+1}]$ denotes the vector of unknown parameters and the superscript T above denotes transpose. The vector $[\mathbf{U}(\mathbf{P})\text{-}\mathbf{G}]^T$ is given by

$$[\mathbf{U}(\mathbf{P})\text{-}\mathbf{G}]^T = [u(x_0, t_1, \mathbf{p}) - g(t_1), u(x_0, t_2, \mathbf{p}) - g(t_2), \cdots, u(x_0, t_n, \mathbf{p}) - g(t_n)]. \tag{9}$$

$F(\mathbf{P})$ is real-valued bounded function defined on a closed bounded domain $D \subset R^{m+1}$. The function $F(\mathbf{P})$ may have many local minimum in D, but it has only one global minimum. When $F(\mathbf{P})$ and D have some attractive properties, for instance, $F(\mathbf{P})$ is a differentiable concave function and D is a convex region, then a local maximum and problem can be solved explicitly by mathematical programming methods.

3. Overview of the Levenberg-Marquardt method

The Levenberg-Marquardt method, originally devised for application to nonlinear parameter estimation problems, has also been successfully applied to the solution of linear ill-conditioned problems. Such a method was first derived by Levenberg (1944) by modifying the ordinary least-squares norm. Later Marquardt (1963) derived basically the same technique by using a different approach. Marquardt's intention was to obtain a method that would tend to the Gauss method in the neighborhood of the minimum of the ordinary least-squares norm, and would tend to the steepest descent method in the neighborhood of the initial guess used for the iterative procedure.

To minimize the least squares norm (8), we need to equate to zero the derivatives of $F(\mathbf{P})$ with respect to each of the unknown parameters $[p_1, p_2, \cdots, p_{m+1}]$, that is

$$\frac{\partial F(\mathbf{P})}{\partial p_1} = \frac{\partial F(\mathbf{P})}{\partial p_2} = \cdots = \frac{\partial F(\mathbf{p})}{\partial p_{m+1}} = 0. \tag{10}$$

Let us introduce the sensitivity or Jacobian matrix, as follows:

$$\mathbf{J}(\mathbf{P}) = \left[\frac{\partial \mathbf{U}^T(\mathbf{P})}{\partial \mathbf{P}}\right]^T = \begin{bmatrix} u_{p_1}(x_0, t_1, \mathbf{p}) & u_{p_2}(x_0, t_1, \mathbf{p}) & \cdots & u_{p_{m+1}}(x_0, t_1, \mathbf{p}) \\ u_{p_1}(x_0, t_2, \mathbf{p}) & u_{p_2}(x_0, t_2, \mathbf{p}) & \cdots & u_{p_{m+1}}(x_0, t_2, \mathbf{p}) \\ \cdots & & & \\ u_{p_1}(x_0, t_n, \mathbf{p}) & u_{p_2}(x_0, t_n, \mathbf{p}) & \cdots & u_{p_{m+1}}(x_0, t_n, \mathbf{p}) \end{bmatrix}, \tag{11}$$

$$\text{or} \quad J_{ij} = u_{p_j}(x_0, t_i, \mathbf{p}) = \frac{\partial u(x_0, t_i, \mathbf{p})}{\partial p_j}, i = 1, 2, \ldots n, j = 1, 2, \ldots m + 1. \tag{12}$$

The elements of the sensitivity matrix are called the sensitivity coefficients, the results of differentiation (10) can be written down as follows:

$$-2\mathbf{J}^T(\mathbf{P})[\mathbf{U}(\mathbf{P}) - \mathbf{G}] = 0. \tag{13}$$

For linear inverse problem the sensitivity matrix is not a function of the unknown parameters. The Eq. (13) can be solved then in explicit form:

$$\mathbf{P} = (\mathbf{J}^T\mathbf{J})^{-1}\mathbf{J}^T\mathbf{G}. \tag{14}$$

In the case of a nonlinear inverse problem, the matrix **J** has some functional dependence on the vector **p**. The solution of Eq. (13) requires an iterative procedure, which is obtained by linearizing the vector $\mathbf{U}(\mathbf{P})$ with a Taylor series expansion around the current solution at iteration k. Such a linearization is given by

$$\mathbf{U}(\mathbf{P}) = \mathbf{U}\left(\mathbf{P}^k\right) + \mathbf{J}^k\left(\mathbf{P}\text{-}\mathbf{P}^k\right), \tag{15}$$

where $\mathbf{U}\left(\mathbf{P}^k\right)$ and $\mathbf{J}^k$ are the estimated temperatures and the sensitivity matrix evaluated at iteration k, respectively. Eq. (15) is substituted into (14) and the resulting expression is rearranged to yield the following iterative procedure to obtain the vector of unknown parameters **P**:

$$\mathbf{P}^{k+1} = \mathbf{P}^k + \left[\left(\mathbf{J}^k\right)^T\mathbf{J}^k\right]^{-1}\left(\mathbf{J}^k\right)^T\left[\mathbf{G}\text{-}\mathbf{U}\left(\mathbf{P}^k\right)\right]. \tag{16}$$

The iterative procedure given by Eq. (16) is called the Gauss method. Such method is actually an approximation for the Newton (or Newton-Raphson) method. We note that Eq. (14), as well as the implementation of the iterative procedure given by Eq. (16), require the matrix $\mathbf{J}^T\mathbf{J}$ to be nonsingular, or

$$\left|\mathbf{J}^T\mathbf{J}\right| \neq 0, \tag{17}$$

where $|\cdot|$ is the determinant.

Formula (17) gives the so called identifiability condition, that is, if the determinant of $\mathbf{J}^T\mathbf{J}$ is zero, or even very small, the parameters p_j, for $j = 1, 2, \cdots, \mathrm{m} + 1$, cannot be determined by using the iterative procedure of Eq. (16).

Problems satisfying $\left|\mathbf{J}^T\mathbf{J}\right| \approx 0$ are denoted ill-conditioned. Inverse heat transfer problems are generally very ill-conditioned, especially near the initial guess used for the unknown parameters, creating difficulties in the application of Eqs. (14) or (16). The Levenberg-Marquardt method alleviates such difficulties by utilizing an iterative procedure in the form:

$$\mathbf{P}^{k+1} = \mathbf{P}^k + \left[\left(\mathbf{J}^k\right)^T\mathbf{J}^k + \mu^k\Omega^k\right]^{-1}\left(\mathbf{J}^k\right)^T\left[\mathbf{G}\text{-}\mathbf{U}\left(\mathbf{P}^k\right)\right], \tag{18}$$

where μ^k is a positive scalar named damping parameter and Ω^k is a diagonal matrix.

The purpose of the matrix term $\mu^k\Omega^k$ is to damp oscillations and instabilities due to the ill-conditioned character of the problem, by making its components large as compared to those of $\mathbf{J}^T\mathbf{J}$ if necessary. μ^k is made large in the beginning of the iterations, since the problem is generally ill-conditioned in the region around the initial guess used for iterative procedure, which can be quite far from the exact parameters. With such an approach, the matrix $\mathbf{J}^T\mathbf{J}$ is not required to be nonsingular in the beginning of iterations and the Levenberg-Marquardt method tends to the steepest descent method, that is, a very small step is taken in the negative gradient direction. The parameter μ^k is then gradually reduced as the iteration procedure advances to the solution of the parameter estimation problem, and then the Levenberg-Marquardt method tends to the Gauss method given by (16). The following criteria were suggested in literature [13] to stop the iterative procedure of the Levenberg-Marquardt method given by Eq. (18):

$$F\left(\mathbf{p}^{k+1}\right) < \varepsilon_1, \tag{19}$$

$$\left\| \left(\mathbf{J}^k\right) \left[\mathbf{G}\text{-}\mathbf{U}\left(\mathbf{p}^k\right)\right] \right\| < \varepsilon_2, \tag{20}$$

$$\left\| \mathbf{p}^{k+1}\text{-}\mathbf{p}^k \right\| < \varepsilon_3, \tag{21}$$

where ε_1, ε_2 and ε_3 are user prescribed tolerances and $\|\cdot\|$ denotes the Euclidean norm. The criterion given by Eq. (19) tests if the least squares norm is sufficiently small, which is expected in the neighborhood of the solution for the problem. Similarly, Eq. (20) checks if the norm of the gradient of $F(\mathbf{p})$ is sufficiently small, since it is expected to vanish at the point where $F(\mathbf{p})$ is minimum. The last criterion given by Eq. (21) results from the fact that changes in the vector of parameters are very small when the method has converged. Generally, these three stopping criteria need to be tested and the iterative procedure of the Levenberg-Marquardt method is stopped if any of them is satisfied.

Different versions of the Levenberg-Marquardt method can be found in the literature, depending on the choice of the diagonal matrix Ω^k and on the form chosen for the variation of the damping parameter μ^k. In this paper, we choose the Ω^k as

$$\Omega^k = diag\left[\left(\mathbf{J}^k\right)^T \mathbf{J}^k\right]. \tag{22}$$

Suppose that the vector of temperature measurements $\mathbf{G} = [\mathrm{g}(\mathrm{t}_1), \mathrm{g}(\mathrm{t}_2), \cdots, \mathrm{g}(\mathrm{t}_n)]$ are given at times t_i, $i = 1, 2, \cdots, \mathrm{n}$ and an initial guess $\mathbf{P}^0$ is available for the vector of unknown parameters $\mathbf{P}$. Choose a value for μ^0, say, $\mu^0 = 0.001$ and $k = 0$. Then,

Step 1. Solve the direct problem (1)–(4) with the available estimate $\mathbf{P}^k$ in order to obtain the vector $\mathbf{U}\left(\mathbf{P}^k\right) = \left[u\left(x_0, t_1, \mathbf{p}^k\right), u\left(x_0, t_2, \mathbf{p}^k\right), \cdots, u\left(x_0, t_n, \mathbf{p}^k\right)\right]$.

Step 2. Compute $F\left(\mathbf{P}^k\right)$ from the Eq. (8).

Step 3. Compute the sensitivity matrix $\mathbf{J}^k$ from (12) and then the matrix Ω^k from (22), by using the current value of $\mathbf{P}^k$.

Step 4. Solve the following linear system of algebraic equations, obtained from (18): $\left[\left(\mathbf{J}^k\right)^T \mathbf{J}^k + \mu^k \Omega^k\right] \Delta\mathbf{P}^k = \left(\mathbf{J}^k\right)^T \left[\mathbf{G}\text{-}\mathbf{U}\left(\mathbf{P}^k\right)\right]$ in order to compute $\Delta\mathbf{P}^k = \mathbf{P}^{k+1}\text{-}\mathbf{P}^k$.

Step 5. Compute the new estimate $\mathbf{P}^{k+1}$ as $\mathbf{P}^{k+1} = \mathbf{P}^k + \Delta\mathbf{P}^k$.

Step 6. Solve the exact problem (1)–(4) with the new estimate $\mathbf{P}^{k+1}$ in order to find $U\left(\mathbf{P}^{k+1}\right)$. Then compute $F\left(\mathbf{P}^{k+1}\right)$.

Step 7. If $F\left(\mathbf{P}^{k+1}\right) \geq F\left(\mathbf{P}^k\right)$ replace μ^k by $10\mu^k$ and return to step 4.

Step 8. If $F\left(\mathbf{P}^{k+1}\right) \leq F\left(\mathbf{P}^k\right)$, accept the new estimate $\mathbf{P}^{k+1}$ and replace μ^k by $0.1\mu^k$.

Step 9. Check the stopping criteria given by (19). Stop the iterative procedure if any of them is satisfied; otherwise, replace k by $k + 1$ and return to step 3.

4. Calculation of sensitivity coefficients

Generally, there have two approaches for determining the gradient; the first is a discretize-then-differentiate approach and the second is a differentiate-then-discretize approach.

The first approach is to approximate the gradient of the functional by a finite difference quotient approximation, but in general, we cannot determine the sensitivities exactly, so this method may led to larger error.

Here we intend to use differentiate-then-discretize approach which we refer to as the sensitivity equation method. This method can be determined more efficiently with the help of the sensitivities

$$\mathrm{u}_k = \frac{\partial u}{\partial p_k}, k = 1, 2, \cdots m + 1. \tag{23}$$

We first differentiate the flow system (1)–(4) with respect to each of the design parameters $[p_1, p_2, \ldots, p_{m+1}]$, to obtain the $m + 1$ continuous sensitivity systems: for $k = 1, 2, \cdots, m + 1$

$$\begin{cases} \dfrac{\partial u_k}{\partial t} = \dfrac{\partial}{\partial x}\left[\left(p_1 + p_2 x + \cdots + p_{m+1} x^m\right)\dfrac{\partial u_k}{\partial x} + x^{k-1}\dfrac{\partial u}{\partial x}\right] \\ u_k(x, 0) = 0 \\ u_k(0, t) = 0 \\ u_k(L, t) = 0 \end{cases}. \tag{24}$$

There have $(m + 2)$ equations, we can make them in one system equation and use the finite element methods to solve the system of equation. Here, we give the vector form of the equation as follow:

$$(\mathbf{P1})\begin{cases} \dfrac{\partial \vec{\mathbf{U}}}{\partial t} + \nabla \cdot \vec{\mathbf{\Gamma}} = \vec{\mathbf{F}} \\ \vec{\mathbf{U}}(x, 0) = \vec{\mathbf{U}}_0(x), \\ \vec{\mathbf{U}}(0, t) = \vec{\mathbf{G}}_1(t) \\ \vec{\mathbf{U}}(L, t) = \vec{\mathbf{G}}_2(t) \end{cases} \tag{25}$$

where

$$\vec{\mathbf{U}} = \begin{bmatrix} u \\ u_1 \\ u_2 \\ \cdots \\ u_{m+1} \end{bmatrix}, \vec{\mathbf{\Gamma}} = \begin{bmatrix} -\left(p_1 + p_2 x + \cdots + p_{m+1} x^m\right)\dfrac{\partial u}{\partial x} \\ -\left(p_1 + p_2 x + \cdots + p_{m+1} x^m\right)\dfrac{\partial u_1}{\partial x} - \left(\dfrac{\partial u}{\partial x}\right) \\ -\left(p_1 + p_2 x + \cdots + p_{m+1} x^m\right)\dfrac{\partial u_2}{\partial x} - x\left(\dfrac{\partial u}{\partial x}\right) \\ \cdots \\ -\left(p_1 + p_2 x + \cdots + p_{m+1} x^m\right)\dfrac{\partial u_{m+1}}{\partial x} - \left(x^m \dfrac{\partial u}{\partial x}\right) \end{bmatrix}, \vec{\mathbf{F}} = \begin{bmatrix} f(\mathrm{x}, \mathrm{t}) \\ 0 \\ 0 \\ \cdots \\ 0 \end{bmatrix}. \tag{26}$$

$$\vec{\mathbf{U}}_0(x) = [\mathrm{u}_0(x), 0, 0, \cdots 0]^T, \vec{\mathbf{G}}_1(t) = [g_1(t), 0, 0, \cdots 0]^T, \vec{\mathbf{G}}_2(t) = [g_2(t), 0, 0, \cdots 0]^T. \tag{27}$$

We use the Galerkin finite element method approximation for discretizing problem (25). For this, we multiply the Eq. (25) by a test function $v : [0, L] \to R$, $v \in V_0 := H_0^1(0, L)$ and integrate the obtained equation in space form 0 to L. We obtain the following equation:

$$\int_0^L \frac{\partial \mathbf{U}(x, t)}{\partial t} \cdot v(x)dx \text{-} \int_0^L \nabla \mathbf{\Gamma} \cdot v(x)dx = \int_0^L \mathbf{F}(x, t) \cdot v(x)dx, \tag{28}$$

integrating by parts gives

$$\int_0^L \nabla \mathbf{\Gamma} \cdot v(x)dx = (\mathbf{\Gamma} \cdot v(x))|_0^L - \int_0^L \mathbf{\Gamma} \cdot \frac{\partial v(x, t)}{\partial x} dx. \tag{29}$$

We can change the first derivative in time and the integral. We have $v(0) = 0 = v(L)$, because $v \in V_0$. This leads to an equivalent problem to (**P1**): $\forall t>0$, find $\mathbf{U}(x,t)$ satisfying

$$\frac{d}{dt}\int_0^L \mathbf{U}(x,t)\cdot v(x)dx + \int_0^L \mathbf{\Gamma}\cdot\frac{\partial v(x,t)}{\partial x}dx = \int_0^L \mathbf{F}(x,t)\cdot v(x)dx, \tag{30}$$

for all $v \in V_0 := H_0^1(0,L)$. To simplify the notation we use the scalar product in $L^2(0,L)$

$$(f,g) = \int_0^L f(x)\cdot g(x)dx. \tag{31}$$

We also can define the following bilinear form:

$$a(\mathbf{U},v) = \int_0^L \mathbf{\Gamma}\cdot\frac{\partial v(x,t)}{\partial x}dx = \begin{cases} \int_0^L -(p_1+p_2x+\cdots+p_{m+1}x^m)\frac{\partial u}{\partial x}\frac{\partial v}{\partial x}dx \\ \int_0^L\left(-(p_1+p_2x+\cdots+p_{m+1}x^m)\frac{\partial u_1}{\partial x}\frac{\partial v}{\partial x}-\frac{\partial u}{\partial x}\frac{\partial v}{\partial x}\right)dx \\ \int_0^L\left(-(p_1+p_2x+\cdots+p_{m+1}x^m)\frac{\partial u_2}{\partial x}\frac{\partial v}{\partial x}-x\frac{\partial u}{\partial x}\frac{\partial v}{\partial x}\right)dx \\ \cdots \\ \int_0^L\left(-(p_1+p_2x+\cdots+p_{m+1}x^m)\frac{\partial u_{m+1}}{\partial x}\frac{\partial v}{\partial x}-x^{m+1}\frac{\partial u}{\partial x}\frac{\partial v}{\partial x}\right)dx \end{cases}. \tag{32}$$

Finally, we obtain with this notations the weak problem of (**P1**):

$$(\mathbf{P2})\begin{cases} \frac{d}{dt}(\mathbf{U},v)_{L^2} + a(\mathbf{U},v) = (\mathbf{F},v)_{L^2} \\ \mathbf{U}(x,0) = \mathbf{U}_0(x) \\ \mathbf{U}(0,t) = \mathbf{G}_1(t) \\ \mathbf{U}(L,t) = \mathbf{G}_2(t) \end{cases}. \tag{33}$$

4.1 Space-discretization with the Galerkin method

In this section, we search a semi-discrete approximation of the weak problem (**P2**), using the Galerkin finite element method. This leads to a first order Cauchy-problem in time.

Let V_h be a $N_x + 1$ dimensional subspace of V and $V_{0,h} = V_h \cap V_0$. Then the following problem is an approximation of the weak problem, find $u_h, u_{1,h}, u_{2,h}\cdots u_{m+1.h} \in V_h$ that satisfies:

$$\begin{cases} \frac{d}{dt}(\mathbf{U}_h,v_h) + a(\mathbf{U}_h,v_h) = (\mathbf{F},v_h) \\ \mathbf{U}_h(\mathbf{x},0) = \mathbf{U}_{0,h}(x) \\ \mathbf{U}_h(0,t) = \mathbf{G}_1(t) \\ \mathbf{U}_h(L,t) = \mathbf{G}_2(t) \end{cases}, \tag{34}$$

for all $v_h \in V_{0,h}$. where $\mathbf{U}_h = [u_h, u_{1,h}, u_{2,h}\cdots u_{m+1.h}]^T$.

The choice of V_h is completely arbitrary. So we can choose it the way that for later treatment, it will be as easy as possible. For example, we subdivide the interval $[0, L]$ into partitions of equal distances h:

$$0 = a_1 < a_2 < \cdots < a_{N_x} < a_{N_x+1} = L, , a_i = (i-1) \cdot h \tag{35}$$

$$V_h = \left\{ v_h \in C^0[0, L] : v_h \big|_{[a_i, a_{i+1}]} \in \mathrm{P}_1, \forall i = 1 \cdots N_x \right\}, \tag{36}$$

$$V_{0,h} = \{ v_h \in V_h : v_h(0) = v_h(L) = 0 \}. \tag{37}$$

Note, that the finite dimension allows us to build a finite base for the corresponding space. In the case of $V_{0,h}$, we have: $\{\varphi_i\}_{i=2}^{N_x}$ where $\forall i = 2 \cdots N_x$.

$$\varphi_i(x) = \begin{cases} 0 & x \in [a_0, a_{i-1}] \\ \dfrac{x}{h} - (i-2) & x \in [a_{i-1}, a_i] \\ i - \dfrac{x}{h} & x \in [a_i, a_{i+1}] \\ 0 & x \in [a_{i+1}, a_{N_x+1}] \end{cases}, \tag{38}$$

while we add for V_h the two functions φ_1 and φ_{N_x+1} defined as:

$$\varphi_1(x) = \begin{cases} 1 - \dfrac{x}{h}, & if x \in [a_1, a_2] \\ 0, & if x \in [a_2, a_{N_x+1}] \end{cases}, \tag{39}$$

$$\varphi_{N_x+1}(x) = \begin{cases} 0, & if x \in [a_1, a_{N_x}] \\ \dfrac{x}{h} - N_x + 1, & if x \in [a_{N_x}, a_{N_x+1}] \end{cases}, \tag{40}$$

so that we can write $\mathbf{U}_h$ as a linear combination of the basic elements:

$$\mathbf{U}_h(x, t) = \sum_{j=1}^{N_x+1} \tilde{\mathbf{U}}_j(t) \cdot \varphi_j(x), \tag{41}$$

$$\mathbf{U}_{0,h}(x, t) = \sum_{j=1}^{N_x+1} \mathbf{U}_0(x_j) \cdot \varphi_j(x), \tag{42}$$

where $\tilde{\mathbf{U}}_1(t) = \mathbf{G}_0(t)$ and $\tilde{\mathbf{U}}_{N_x+1}(t) = \mathbf{G}_1(t)$. Using that $a(\cdot, \cdot)$ is bilinear form and that Eq. (34) is valid for each element of the base $\{\varphi_i\}_{i=2}^{N_x}$, we obtain

$$\sum_{j=1}^{N_x+1} \frac{d}{dt} \tilde{\mathbf{U}}_j(t) \cdot \left(\varphi_j, \varphi_i \right) + \sum_{j=1}^{N_x+1} \tilde{\mathbf{U}}_j(t) \cdot a\left(\varphi_j, \varphi_i \right) = (\mathbf{F}, \varphi_i), \tag{43}$$

$\forall i = 2 \cdots N_x$. This equation can be written in a vector form. For this we define the vectors $\vec{\mathbf{u}}$, $\vec{\mathbf{u}}_0$ and $\vec{\mathbf{F}}$ with components

$$\mathbf{F}_i(t) = (\mathbf{F}, \varphi_i)_{L^2}, \mathbf{u}_j(t) := \tilde{\mathbf{u}}_{\mathbf{j}}(t), \mathbf{u}_{0\,,\,\mathbf{j}} = \mathbf{u}_0(\mathbf{x}_{\mathbf{j}}), \tag{44}$$

and matrices $\mathbf{M}$ and $\mathbf{A}$ as

$$m_{ij} := \left(\varphi_i, \varphi_j \right)_{L^2}, a_{ij} := a\left(\varphi_i, \varphi_j \right), \tag{45}$$

Note that $\mathbf{M}, \mathbf{A} \in \mathbf{R}^{N_x - 1 \times N_x + 1}, \vec{\mathbf{u}} \in R^{N_x+1}, and\ \vec{\mathbf{F}} \in \mathbf{R}^{N_x - 1}$. So that (43) is equal to the Cauchy problem

$$\begin{cases} \mathbf{M}\frac{d}{dx}\vec{\mathbf{u}}(t) + \mathbf{A} \cdot \vec{\mathbf{u}}(t) = \vec{\mathbf{F}}(t) \\ \vec{\mathbf{u}}(t_0) = \vec{\mathbf{u}}_\mathbf{0} \end{cases}, \tag{46}$$

the Crank-Nicolson method can be applied to (46) at time t_k, resulting in

$$\mathbf{M} \cdot \left(\frac{\vec{\mathbf{u}}_{k+1} - \vec{\mathbf{u}}_k}{\Delta t}\right) + \frac{1}{2}\mathbf{A} \cdot \vec{\mathbf{u}}_{\mathbf{k+1}} + \frac{1}{2}\mathbf{A} \cdot \vec{\mathbf{u}}_\mathbf{k} = \frac{1}{2}\left(\vec{\mathbf{F}}_\mathbf{k} + \vec{\mathbf{F}}_{\mathbf{k+1}}\right), \tag{47}$$

where $\vec{\mathbf{u}}_\mathbf{k} = \vec{\mathbf{u}}(t_k)$, $\vec{\mathbf{F}}_\mathbf{k} = \vec{\mathbf{F}}(t_k), k = 0, 1, \cdots$..
The Eq. (47) can be written in simple form as

$$\left(\mathbf{M} + \frac{\Delta t}{2}\mathbf{A}\right) \cdot \vec{\mathbf{u}}_{\mathbf{k+1}} = \left(\mathbf{M} - \frac{\Delta t}{2}\mathbf{A}\right) \cdot \vec{\mathbf{u}}_\mathbf{k} + \frac{\Delta t}{2}\left(\vec{\mathbf{F}}_\mathbf{k} + \vec{\mathbf{F}}_{\mathbf{k+1}}\right), \tag{48}$$

the algebraic system (48) is solved by Gauss elimination method.

5. Numerical experiment

In this section, we are going to demonstrate some numerical results for $(\mathrm{u}(\mathrm{x}, \mathrm{t}), \mathrm{q}(\mathrm{x}))$ in the inverse problem (1)–(5). Therefore the following examples are considered and the solution is obtained.

Example 1. Consider (1)–(4) with

$$u(\mathrm{x}, 0) = \mathrm{sinx}, \quad 0 \le x \le 1, \tag{49}$$

$$u(0, \mathrm{t}) = 0, \quad 0 \le t \le 1 \tag{50}$$

$$u(1, \mathrm{t}) = \sin(1)\mathrm{e}^{-t}, \quad 0 \le t \le 1, \tag{51}$$

$$f(x,t) = \left(\sin x\left(\frac{x^2}{4} + \frac{x}{2} + 1\right)\right)e^{-t} - \frac{(x+1)}{2}\cos(x)e^{-t} - \sin x e^{-t}, \quad 0 \le x \le 10 \le t \le 1 \tag{52}$$

We obtain the unique exact solution

$$q(\mathrm{x}) = 1 + 0.5\mathrm{x} + 0.25\mathrm{x}^2 \tag{53}$$

And

$$u(\mathrm{x}, \mathrm{t}) = \sin(\mathrm{x})\mathrm{e}^{-t}. \tag{54}$$

We take the observed data g as

$$g(t) = u(0.5, \mathrm{t}) = \sin(0.5)\mathrm{e}^{-t} 0 \le t \le 1 \tag{55}$$

The unknown function $q(\mathrm{x})$ defined as the following form

$$q(\hat{\mathrm{x}}) = p_1 + p_2 x + p_3 x^2, \tag{56}$$

where p_1, p_2, p_3 are unknown coefficients.

Table 1 shows how the Levenberg-Marquardt algorithm can find the best parameters after 12 iterations when it is initialized in four different points.

Figures 1–4 show the fitness of the estimated parameters and the rate of convergence.

Figures 5–8 show the comparison between the inversion results $\hat{q}(x)$ and the exact value $q(x)$:

Table 2 shows the values of $q(\mathrm{j}\Delta\mathrm{x})$ and $u(\mathrm{j}\Delta\mathrm{x}, 0.5)$ in $x = j\Delta x$ with the all the initial values are set 1.

Example 2. Consider (1)–(4) with

$$u(\mathrm{x}, 0) = \mathrm{xe}^{-x}, \quad 0 \leq x \leq 1, \tag{57}$$

$$u(0, \mathrm{t}) = \mathrm{te}^{-t}, \quad 0 \leq t \leq 1 \tag{58}$$

$$u(1, \mathrm{t}) = (1 + \mathrm{t})\mathrm{e}^{-1-t}, \quad 0 \leq t \leq 1 \tag{59}$$

$$f(\mathrm{x}, \mathrm{t}) = \mathrm{e}^{-(\mathrm{t+x})} \text{-} (\mathrm{t} + \mathrm{x})\mathrm{e}^{-(\mathrm{t+x})} \text{-} e^{x}\left(\mathrm{e}^{-(\mathrm{t+x})} \text{-} (\mathrm{t} + \mathrm{x})\mathrm{e}^{-(\mathrm{t+x})}\right) + \mathrm{e}^{x}\left(2\mathrm{e}^{-(\mathrm{t+x})} \text{-} (\mathrm{t} + \mathrm{x})\mathrm{e}^{-(\mathrm{t+x})}\right), \quad 0 \leq x \leq 10 \leq t \leq 1 \tag{60}$$

Starting point	0.5 0.5 0.5	1 1 1	10 10 10	50 50 50
Iteration 12	0.999729028233135 0.499885876453067 0.252009862457275	0.999729028233183 0.499885876453056 0.252009862457315	0.999729028233194 0.499885876453057 0.252009862457325	0.999729028307261 0.499885876454169 0.25200986249336
Error F	$8.7564944405 \times 10^{-14}$	$8.7564944427 \times 10^{-14}$	$8.7564944420 \times 10^{-14}$	$8.7564944420 \times 10^{-14}$

Table 1.
Performance of the algorithm when it is run to solve the model using three different parameters guesses.

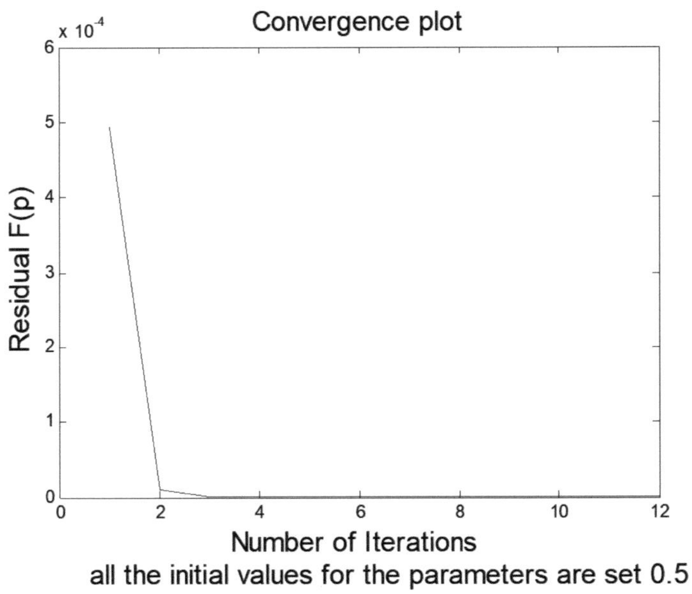

Figure 1.
All the initial values for the parameters are set 0.5.

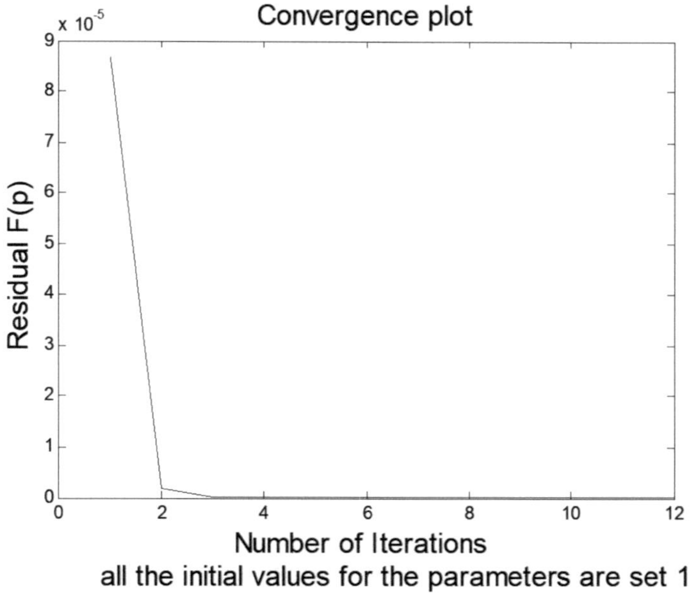

Figure 2.
All the initial values for the parameters are set 1.

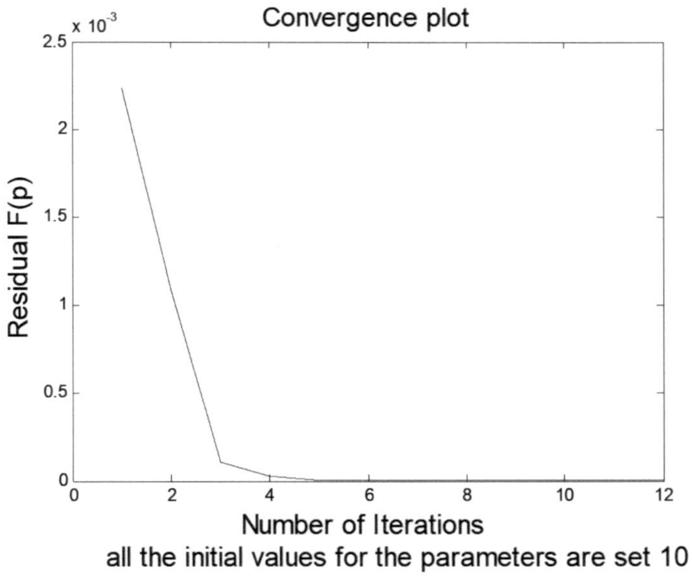

Figure 3.
All the initial values for the parameters are set 10.

We obtain the unique exact solution

$$q(\mathrm{x}) = e^{x}, \tag{61}$$

And

$$u(\mathrm{x},\mathrm{t}) = (\mathrm{x}+\mathrm{t})\mathrm{e}^{-x-t}. \tag{62}$$

We take the observed data g as

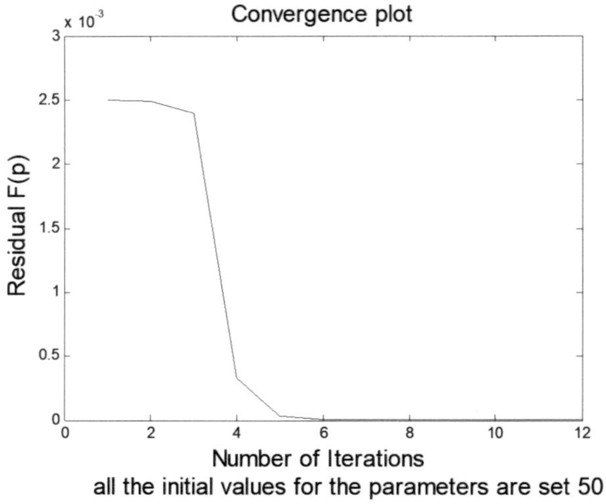

Figure 4.
All the initial values for the parameters are set 50.

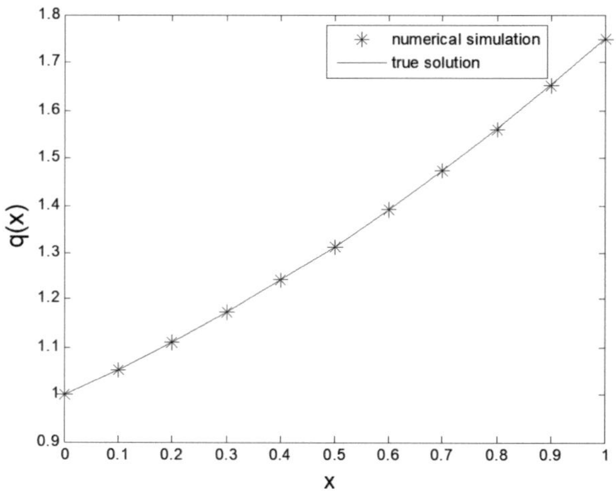

Figure 5.
The comparison chart with all the initial values for the parameters is set 0.5.

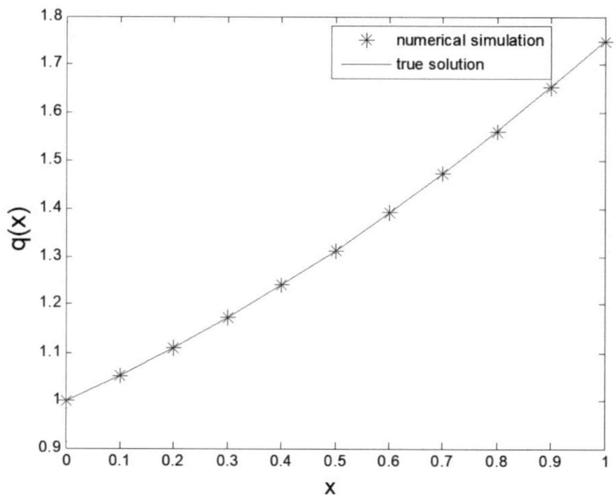

Figure 6.
The comparison chart with all the initial values for the parameters is set 1.

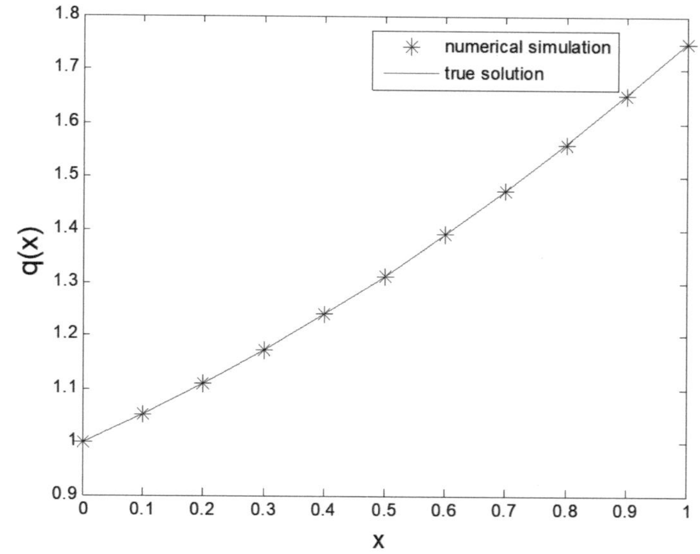

Figure 7.
The comparison chart with all the initial values for the parameters is set 10.

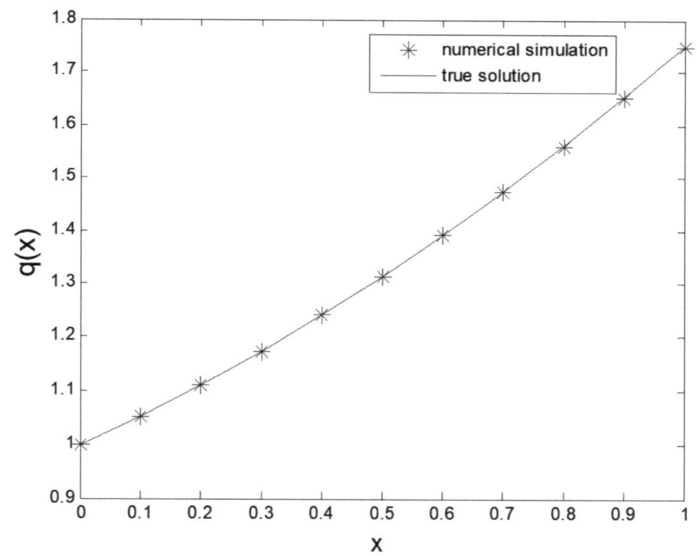

Figure 8.
The comparison chart with all the initial values for the parameters is set 50.

$$g(t) = u(0.5, \mathrm{t}) = (0.5 + \mathrm{t})\mathrm{e}^{(-0.5-\mathrm{t})} 0 \le t \le 1. \tag{63}$$

The unknown function $q(\mathrm{x})$ defined as the following form $\hat{q(\mathrm{x})} = p_1 + p_2x + p_3x^2 + p_4x^3 + p_5x^4 + p_6x^5 + p_7x^6 + p_8x^7$, where $p_1, p_2, \cdots, p_7, p_8$ are unknown coefficients.

Table 3 shows how the Levenberg-Marquardt algorithm can find the best parameters after 20 iterations when it is initialized in four different points.

Figures 9–12 show the fitness of the estimated parameters and the rate of convergence.

Figures 13–16 show the comparison between the inversion results $\hat{q}(x)$ and the exact value $q(x)$:

	Numerical	Exact	Numerical	Exact
j	$q(j\Delta x)$	$q(j\Delta x)$	$u(j\Delta x, 0.5)$	$u(j\Delta x, 0.5)$
0	0.999729028233183	1	0	0
1	1.05223771450306	1.0525	0.0605593190239173	0.0605520280601669
2	1.10978659802209	1.11	0.120511797611786	0.120499040271796
3	1.17237567879026	1.1725	0.179257059078521	0.179242065904716
4	1.24000495680758	1.24	0.236207449080344	0.236194164064666
5	1.31267443207404	1.3125	0.290793943250869	0.290786288212692
6	1.39038410458965	1.39	0.342471828361625	0.342472971890064
7	1.47313397435441	1.4725	0.390726114897089	0.390737778838824
8	1.56092404136831	1.56	0.435076630410587	0.435098463062163
9	1.65375430563136	1.6525	0.475082717530532	0.475111787267016
10	1.75162476714355	1.75	0.510347406713368	0.510377951544573

Table 2.
The values of $q(j\Delta x)$ and $u(j\Delta x, 0.5)$ in $x = j\Delta x$ with the all the initial values being set to 1.

Starting point	**0.1 0.1 0.1 0.1 0.1 0.1 0.1 0.1**	**0.5 0.5 0.5 0.5 0.5 0.5 0.5 0.5**	**1 1 1 1 1 1 1 1**	**2 2 2 2 2 2 2 2**
Iteration 20	1.01536263526644 0.896348846894057 0.954285303464511 −0.890298938193057 1.40131927153131 −0.871276408882294 0.183785623507722 0.0359103726343979	1.01536263500695 0.896348850692318 0.954285278486704 −0.890298849338373 1.40131909032315 −0.871276197318301 0.183785492186491 0.0359104061952515	1.01536263525763 0.896348847022403 0.954285302637587 −0.890298935334171 1.40131926588117 −0.871276402487896 0.18378561965322 0.0359103735933848	1.01536263525905 0.896348846999736 0.954285302790922 −0.890298935876618 1.40131926696099 −0.87127640370648 0.183785620380814 0.0359103734148875
Error *F*	$7.89749200363512 \times 10^{-11}$	$7.89749200363504 \times 10^{-11}$	$7.89749200353888 \times 10^{-11}$	$7.8974920035389 \times 10^{-11}$

Table 3.
Performance of the algorithm when it is run to solve the model using four different parameters guesses.

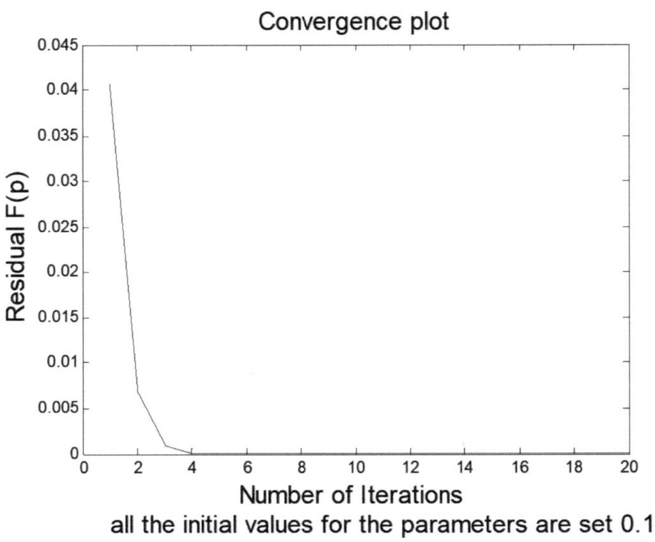

Figure 9.
All the initial values for the parameters are set 0.1.

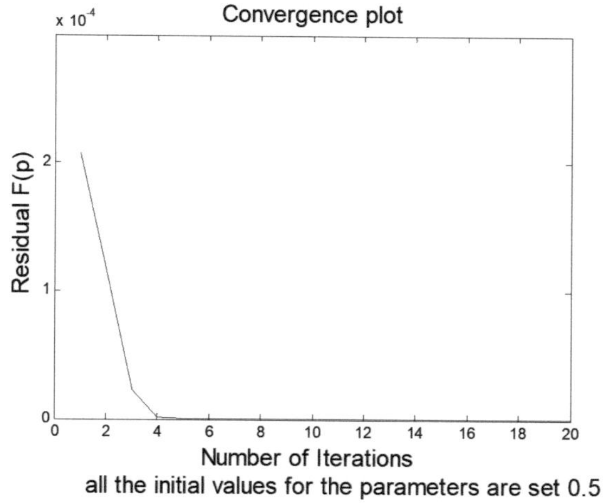

Figure 10.
All the initial values for the parameters are set 0.5.

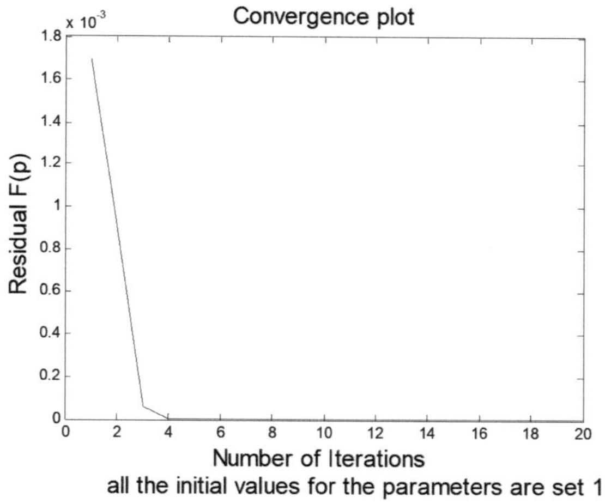

Figure 11.
All the initial values for the parameters are set 1.

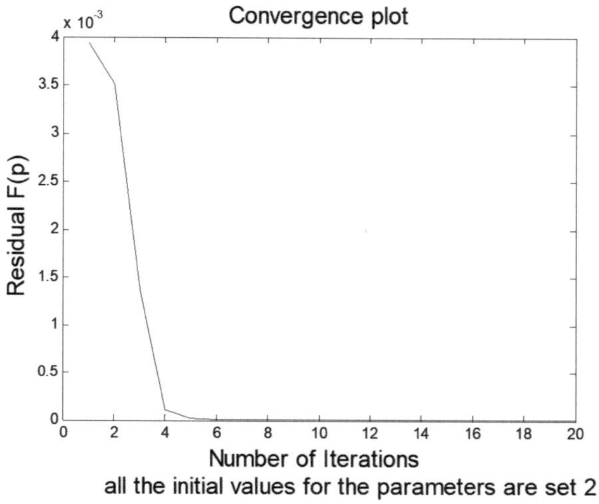

Figure 12.
All the initial values for the parameters are set 2.

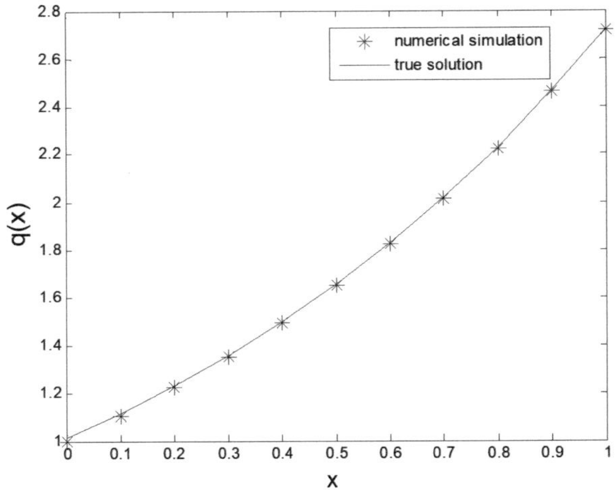

Figure 13.
The comparison chart with all the initial values for the parameters is set 0.1.

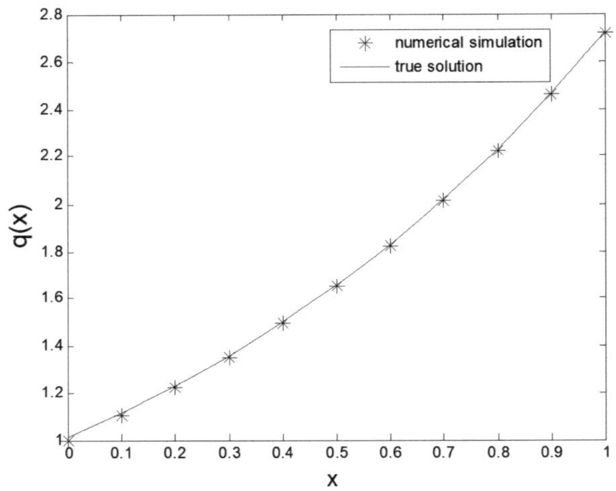

Figure 14.
The comparison chart with all the initial values for the parameters is set 0.5.

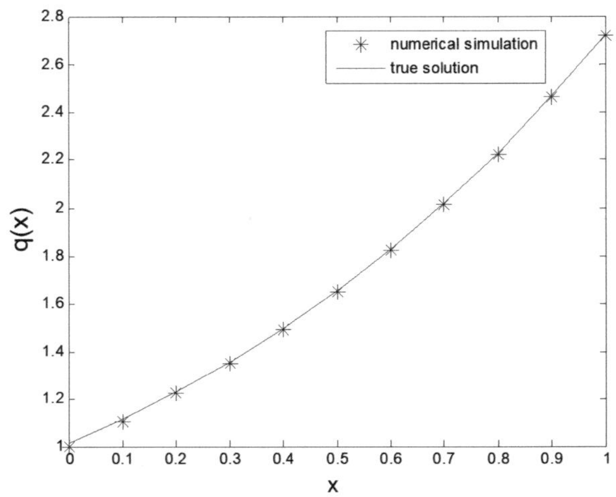

Figure 15.
The comparison chart with all the initial values for the parameters is set 1.

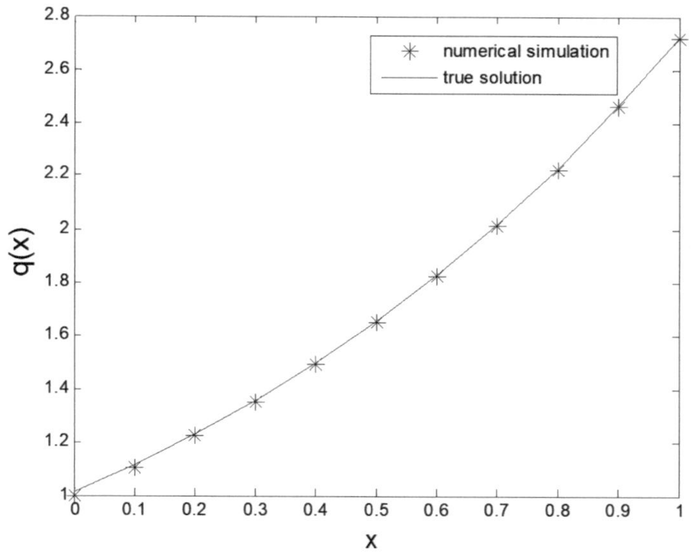

Figure 16.
The comparison chart with all the initial values for the parameters is set 2.

	Numerical	Exact	Numerical	Exact
j	$q(j\Delta x)$	$q(j\Delta x)$	$u(j\Delta x, 0.5)$	$u(j\Delta x, 0.5)$
0	1.01536263525763	1	0.303342962644088	0.303265329856317
1	1.11378168059013	1.10517091807565	0.329201964677126	0.329286981656416
2	1.22765694959399	1.22140275816017	0.347492882224886	0.347609712653987
3	1.35549021305874	1.349858807576	0.359347568702678	0.359463171293777
4	1.4973722149265	1.49182469764127	0.365808711321159	0.365912693766539
5	1.65432828415206	1.64872127070013	0.367792378208857	0.367879441171442
6	1.82785056869393	1.82211880039051	0.366091218454182	0.366158192067887
7	2.01963499046464	2.01375270747048	0.361387761473796	0.361433054294643
8	2.23154102006879	2.22554092849247	0.354267869273063	0.354291330944216
9	2.46579237015687	2.45960311115695	0.345233023618059	0.345235749518249
10	2.72543670622333	2.71828182845905	0.334712604803175	0.334695240222645

Table 4.
The values of $q(j\Delta x)$ and $u(j\Delta x, 0.5)$ in $x = j\Delta x$ with the all the initial values are set 1.

Table 4 shows the values of $q(j\Delta x)$ and $u(j\Delta x, 0.5)$ in $x = j\Delta x$ with the all the initial values are set 1.

6. Conclusions

A numerical method to estimate the temperature $u(x, t)$ and the coefficient $q(x)$ is proposed for an IHCP and the following results are obtained.

1. The present study, successfully applies the numerical method involving the Levenberg-Marquardt algorithm in conjunction with the Galerkin finite element method to an IHCP.

2. From the illustrated example it can be seen that the proposed numerical method is efficient and accurate to estimate the temperature $u(x, t)$ and the coefficient $q(x)$.

Acknowledgements

The work of the author is supported by the Special Funds of the National Natural Science Foundation of China (Nos. 51190093 and 51179151). The author would like to thank the referees for constructive suggestions and comments.

Conflict of interests

The authors declare that there is on conflict of interests regarding the publication of this article.

Author details

Tao Min[1,2*], Xing Chen[1], Yao Sun[1] and Qiang Huang[2]

1 School of Science, Xi'an University of Technology, Xi'an Shaanxi, China

2 State Key Laboratory of Eco-Hydraulic Engineering in Shaanxi, Xi'an University of Technology, Xi'an Shaanxi, China

*Address all correspondence to: mintao@xaut.edu.cn

References

[1] Shidfar A, Karamali GR. Numerical solution of inverse heat conduction problem with nonstationary measurements. Applied Mathematics and Computation. 2005;**168**(1):540-548

[2] Shidfar A, Karamali GR, Damirchi J. An inverse heat conduction problem with a nonlinear source term. Nonlinear Analysis: Theory Methods & Applications. 2006;**65**(3):615-621

[3] Shidfar A, Pourgholi R. Numerical approximation of solution of an inverse heat conduction problem based on Legendre polynomials. Applied Mathematics and Computation. 2006; **175**(2):1366-1374

[4] Shidfar A, Azary H. An inverse problem for a nonlinear diffusion equation. Nonlinear Analysis: Theory Methods & Applications. 1997;**28**(4): 589-593

[5] Kurpisza K, Nowaka AJ. BEM approach to inverse heat conduction problems. Engineering Analysis with Boundary Elements. 1992;**10**(4):291-297

[6] Han H, Ingham DB, Yuan Y. The boundary-element method for the solution of the back ward heat conduction equation. Journal of Computational Physics. 1995;**116**(2): 292-299

[7] Skorek J. Applying the least squares adjustment technique for solving inverse heat conduction problems. In: Taylor C, editor. Proceedings of the 8th Conference on Numerical Methods in Laminar and Turbulent Flow. Swansea: Pineridge Press; 1993. pp. 189-198

[8] Pasquetti R, Le Niliot C. Boundary element approach for inverse heat conduction problems: Application to a bidimensional transient numerical experiment. Numerical Heat Transfer, Part B. 1991;**20**(2):169-189

[9] Ingham DB, Yuan Y. The solution of a nonlinear inverse problem in heat transfer. IMA Journal of Applied Mathematics. 1993;**50**(2):113-132

[10] Hensel E. Inverse Theory and Applications for Engineers. N.J. ISBN: 0135034590: Prentice Hall; 1991

[11] Özisik MN. Inverse Heat Transfer: Fundamentals and Applications. New York, USA: Taylor and Francis. ISBN: 1-56032-838-X; 2000

[12] Pourgholi R, Azizi N, Gasimov YS, Aliev F, Khala HK. Removal of numerical instability in the solution of an inverse heat conduction problem. Communications in Nonlinear Science and Numerical Simulation. 2009;**14**(6): 2664-2669

[13] Shidfar A, Pourgholi R, Ebrahimi M. A numerical method for solving of a nonlinear inverse diffusion problem. Computers and Mathematics with Applications. 2006;**52**(6–7):1021-1030

[14] Wang J, Zabaras N. A Bayesian inference approach to the inverse heat conduction problem. International Journal of Heat and Mass Transfer. 2004;**47**(17–18):3927-3941

[15] Dehghan M. Determination of an unknown parameter in a semi-linear parabolic equation. Mathematical Problems in Engineering. 2002;**8**(2): 111-122

[16] Lihua J, Changfeng M. On L-M method for nonlinear equations. Journal of Mathematics. Wuhan University. 2009;**29**(3):253-259

[17] Chen P. Why not use the Levenberg–Marquardt method for fundamental matrix estimation? The Institution of Engineering and Technology. 2010;**4**(4):286-288

Chapter 7

The Fundamental and Application of Surface Heat Flux Estimation by Inverse Method in Cryogen Spray Cooling

Bin Chen, Jia-Meng Tian and Zhi-Fu Zhou

Abstract

Surface heat flux is an important parameter in various industrial applications, which is often estimated based on measured temperature by solving inverse heat conduction problem (IHCP). In this chapter, the available IHCP methods including sequential function specification (SFS), transfer function (TF) and Duhamel's theorem were compared, taking the example of surface heat flux estimation during spray cooling. The Duhamel's theorem was improved to solve 1D multi-layer ICHP. Considering the significant nonuniformity of heat transfer, the 2D filter solution method was proposed to estimate surface heat flux for 2D multi-layer mediums. The maximum heat flux calculated by the 1D method was underestimated by 60% than that calculated by 2D filter solution, indicating that the lateral heat transfer cannot be ignored. The cooling performances based on 2D filter solution demonstrated that substituting the environment friendly R1234yf for R134a can remarkably reduce global warming potential to <1, but its cooling capacity is insufficient. The effective heat flux of R1234yf can be enhanced by 18.8% by reducing the nozzle diameter and decreasing the back pressure, providing the theoretical basis for the clinical potential substitution of R1234yf with low global warming potential (GWP) for commercial R134a with high GWP in laser dermatology.

Keywords: surface heat flux, inverse heat conduction problem, improved Duhamel's theorem, 2D filter solution, spray cooling

1. Introduction

Heat flux is an important parameter to characterize heat transfer performance in many industrial applications, such as thermal protection of space shuttles [1], thermal management of electronic devices [2], metal heat treatment [3], maintenance of boilers [4] and nuclear reactors [5], spray cooling [6], geophysics [7], etc. Heat flux is often estimated by surface or internal temperature, which is also termed as inverse heat conduction problem (IHCP).

IHCPs are mathematically ill-posed, and a small error in temperature may significantly affect the accuracy of heat flux estimation [8, 9]. Several analytical and numerical methods have been proposed for the solution of IHCPs, such as sequential function specification (SFS) method, Tikhonov regularization (TR) method, transfer function (TF) method, Duhamel's theory, etc. The SFS method is commonly used to solve IHCPs by minimizing the effect of random errors using temperature data at future time steps based on the least square method [10]. The TR method estimates all of the heat flux simultaneously for all time steps and is usually presented as whole time domain form, which often causes heavy computational load [11, 12]. The TF method analogizes the heat conduction problems to dynamic systems, in which heat flux is treated as the input of the system and the temperature profile as the response [13]. This method is simple in concept and one of the most accurate ways of estimating surface heat flux. However, it is relatively difficult to determine the analytic solution of the transfer function for the complex geometry problem. Duhamel's theorem is based on the principle of superposition and assumes that the substrate thermal response at t equals the total sum of what the substrate experienced in small steps prior to t [14]. This method is simple and widely used in the surface heat flux estimation with known surface temperature. However, its assumption that the internal temperature equals to the surface one for indirect temperature measurement often causes significant calculation errors. Thus, Duhamel's theory needs to be improved.

Recently, a filter solution based on TR has attracted the interest of many researchers [9, 15–19], which minimizes the sum of the squares of the errors between estimated and measured temperatures and stabilized by Tikhonov regularization. This solution is expressed in a digital filter form, allowing for a real-time heat flux estimation, and has been used for heat flux estimation of directional flame thermometer [20]. This method demonstrates superiority when solving IHCPs with a complex geometry. However, it can only be used to solve 1D single-layer IHCPs. The multidimensional, multilayer IHCP has yet to be solved.

Although IHCPs have been extensively investigated with regard to various other applications, little work has been conducted related to surface heat flux estimation during cryogen spray cooling (CSC) in laser dermatology. Spray cooling is widely applied in metallurgy, electronics, power plant, and laser dermatology for vascular skin lesions [21–27], because of the superiorities of high power density, ultrafast cooling rate, uniformity of heat removal, and low fluid inventory. In the laser treatment of vascular skin lesions (e.g., port wine stain, PWS), CSC can be implemented to prevent unwanted thermal damage of the epidermis induced by high laser absorption of melanin. Different with traditional steady spray cooling, CSC is a highly transient process with several tens of milliseconds to avoid cold injury. The transient surface heat flux is crucial for cooling performance evaluation, which needs accurate heat flux estimation method and rapid temperature measurement with fast response and small lag, as well as damping of algorithm [28].

In transient CSC, two typical temperature measurements of fine thermocouple (FTC) and thin-film thermocouple (TFTC) are widely used to measure internal and surface temperature. Aguilar et al. [29] used the SFS method to estimate surface heat flux by internal temperature measured by a type-T FTC placed underneath a thin-layer aluminum foil, positioned on the top of epoxy resin surface to provide rapid heat transfer and mechanical support. Zhou et al. [28, 30, 31] measured time-dependent surface temperature by a type-T TFTC with thickness of 2 μm directly deposited onto the epoxy resin surface; this measurement accurately captured the temperature variation owing to its ultrafast thermal response (~1.2 μs). Then, the surface heat flux was estimated by Duhamel's theorem. However, TFTC cannot be used to measure the metal material temperature due to the electrical conductivity. Moreover, TFTC corrodes and oxidizes easily in high-temperature environments. Therefore, FTC measurement is widely used in many industries owing to its reliability and stability. Different with TFTC measurement with single-layer geometry, FTC measurement consists of three layers, namely, aluminum, thermal paste, and epoxy resin. Moreover, the radial and temporal surface temperature variations during CSC result in significant nonuniformity of the surface heat flux [14, 32]. Therefore, lateral heat transfer must be considered. For generality, 2D multilayer IHCPs need to be developed.

In this chapter, the SFS, TF, and Duhamel's theorem methods for TFTC and FTC measurements were compared based on one hypothetical heat flux. Duhamel's theorem was improved to increase the accuracy of surface heat flux when using the indirect three-layer FTC measurement. Afterwards, the 2D filter solution was proposed to calculate the surface heat flux for 2D multilayer mediums. Six hypothetical triangular pulse heat fluxes were used to examine the accuracy and sensitivity of the algorithm. Finally, the 2D filter solution was employed to calculate the surface heat flux to investigate the cooling performance. Using estimated heat flux as evaluation index, the possibility of substituting the commercial cryogen R134a with high GWP (1430) by environment friendly cryogen with low GWP (<1) was discussed.

2. 1D algorithms for heat flux estimation

2.1 Sequential function specification method

The SFS method is widely used to solve IHCPs by minimizing the error between the measured temperature Y_k and estimated temperature T_k for the current time and r future time steps based on the least square method [10]

$$S_M = \sum_{i=1}^{R} \left(Y_{k+r-1} - T_{i,\,k+r-1}\right)^2 \tag{1}$$

Several functional forms of $q(t)$ from t_k to t_{k+r-1} have been proposed. The simplest one is that $q(t)$ is a constant

$$q_k = q_{k+1} = \cdots = q_{k+r} \tag{2}$$

Then, the temperature distribution is represented as a function of surface heat flux q_k, and the temperature field is expanded in a Taylor series about a known value of surface heat flux $\hat{q}_k$ as

$$T_{i,\,k}(q_k) = T_{i,\,k}(\hat{q}_k) + Z_{i,\,k}(q_k - \hat{q}_k) + \cdots \tag{3}$$

where $Z_{i,k}$ is the sensitivity coefficient defined by

$$Z_{i,\,k} \equiv \frac{\partial T_{i,\,k}(q_k)}{\partial q_k} \tag{4}$$

The solution for the estimated surface heat flux at time t_k can be obtained by minimizing Eq. (1) with respect to q_k:

$$q_k = \hat{q}_k + \frac{\sum_{i=1}^{R} \left[Y_{k+r-1} - T_{k+r-1}(\hat{q}_k)\right] Z_{i,\,r-1}^{k}}{\sum_{i=1}^{R} \left(X_{i,\,r-1}^{k}\right)^2} \tag{5}$$

The values of r are selected based on the residual principle [33], which are Eqs. (3) and (4) for the direct and indirect temperature measurement methods, respectively. The detailed information about the SFS method can be found in previous publications [10, 34].

2.2 Transfer function method

The transfer functions establish the relationship between the input and output in a dynamic system, which can also be used to solve the linear heat conduction problems, where the heat flux is treated as the input of the system and the temperature is treated as the response [13].

The estimated surface heat flux using Laplace transform can be described as

$$q(t) = L^{-1}\left[\frac{1}{H_c(s)}\right] * \theta_c(t) \tag{6}$$

where $\theta_c(t)$ is the temperature allowance ($\theta_c(t) = T_c(t) - T_0$), the subscript c denotes the measurement position, the superscript* is the convolution integral operator, and T_0 is the initial temperature. $L^{-1}[1/H(s)]$ is the Laplace inverse transform of the transfer function, which can be written as the function of time t, as follows:

$$L^{-1}\left[\frac{1}{H(s)}\right] = f(t) \tag{7}$$

Substituting Eq. (7) into Eq. (6), $q(t)$ becomes

$$q(t) = \int_0^t \theta_c(t-\tau) f(\tau) \mathrm{d}\tau \tag{8}$$

After dispersing Eq. (8), it takes the following form:

$$\begin{bmatrix} q_1 \\ q_2 \\ \vdots \\ q_n \end{bmatrix} = \begin{bmatrix} \theta_{c,1} & 0 & \cdots & 0 \\ \theta_{c,2} & \theta_{c,1} & \cdots & 0 \\ \vdots & \vdots & \ddots & \vdots \\ \theta_{c,n} & \theta_{c,n-1} & \cdots & \theta_{c,1} \end{bmatrix} \begin{bmatrix} f_0 \\ f_1 \\ \vdots \\ f_n \end{bmatrix} \tag{9}$$

An assist question can be established to solve the temperature allowance θ assuming the boundary condition $q(t) = 1$. Substituting the numerical solution into Eq. (8), the surface heat flux $q(t)$ can be obtained.

2.3 Duhamel's theorem method

Duhamel's theorem directly treated the measured internal temperature as the surface temperature of the substrate when using indirect FTC temperature measurement. A one-dimensional, direct heat conduction problem for the two surface temperature measurements can be solved to calculate the temperature distribution in the substrate [14, 35]:

$$T(x,t) = f(0)\varphi(x,t) + \int_0^t \varphi(x, t-\tau) \frac{dT(\tau)}{d\tau} d\tau \tag{10}$$

where $\varphi(x,t)$ is the temperature distribution response unit step function of the substrate and $f(0)$ and $T(\tau)$ are the initial and time-varying surface temperature. The unit step function for a semi-infinite planar solid is [35].

$$\varphi(x,t) = 1 - erf\left(\frac{x}{2\sqrt{\alpha t}}\right) \tag{11}$$

Solving Eq. (10), the temperature gradient at the surface can be obtained:

$$q(t) = \sqrt{\frac{\lambda \rho c}{\pi}} \int_0^t \frac{1}{\sqrt{t-\tau}} \frac{dT}{d\tau} d\tau \tag{12}$$

The simplification that the internal temperature equals the surface ones ignoring the heat dissipating capacity of the materials above the thermocouple, thereby often causing significant heat flux errors. Therefore, Duhamel's theorem needs to be improved to deal with this problem, rather than directly estimate surface heat flux from the measured temperature data. Firstly, the real surface temperature is needed to be calculated from the internal measured temperature at $x = c$ location, which can be expressed by a piecewise constant function of time as

$$f = f(i \cdot \Delta t), \quad i = 1, 2, \ldots, n, \tag{13}$$

where f is the surface temperature at the time of $i{\cdot}\Delta t$. The temperature at $x = c$ location can be solved by Duhamel's theorem as [35].

$$\begin{aligned} T_c(n \cdot \Delta t) = & f(0)\varphi(x, n \cdot \Delta t) + [f(1 \cdot \Delta t) - f(0)]\varphi(x, (n-1) \cdot \Delta t) \\ & + [f(2 \cdot \Delta t) - f(1 \cdot \Delta t)]\varphi(x, (n-2) \cdot \Delta t) \cdots \\ & + [f(n \cdot \Delta t) - f((n-1) \cdot \Delta t)]\varphi(x, 0) \end{aligned} \tag{14}$$

Eq. (14) can be written in expanded matrix form as

$$\begin{bmatrix} \Delta\varphi(\Delta t) & 0 & \cdots & 0 \\ \Delta\varphi(2\Delta t) & \Delta\varphi(\Delta t) & \cdots & 0 \\ \vdots & \vdots & \ddots & 0 \\ \Delta\varphi(n\Delta t) & \Delta\varphi((n-1)\Delta t) & \cdots & \Delta\varphi(\Delta t) \end{bmatrix} \begin{bmatrix} f(0 \cdot \Delta t) \\ f(1 \cdot \Delta t) \\ \vdots \\ f((n\text{-}1) \cdot \Delta t) \end{bmatrix} = \begin{bmatrix} T_c(0 \cdot \Delta t) \\ T_c(1 \cdot \Delta t) \\ \vdots \\ T_c(n \cdot \Delta t) \end{bmatrix} \tag{15}$$

where $\Delta\varphi(i\Delta t)$ represents $\varphi(x, i{\cdot}\Delta t) - \varphi(x, (i-1){\cdot}\Delta t)$. The surface temperature can be obtained by multiplying the inverse temperature transformation matrix on both sides of the equation from the measured internal temperature data. Afterwards, the surface heat flux can be estimated using Eq. (12).

2.4 Validation of 1D filter solution

Two different methods were employed to measure the surface temperature during CSC. **Figure 1** shows a schematic of the thin-film thermocouple (TFTC) and the fine thermocouple (FTC) measurements. The type-T TFTC with a thickness of 2 μm was directly deposited onto the epoxy resin surface using the magnetron technique. It has perfect contact with the underlying substrate and a fast response time (~1 μs). The FTC measurement with a response time of 3.33 ms [36] consists of a fine type-T thermocouple (~10 μm bead diameter) underneath a thin layer of aluminum foil (~10 μm), which is positioned on the surface of the epoxy resin

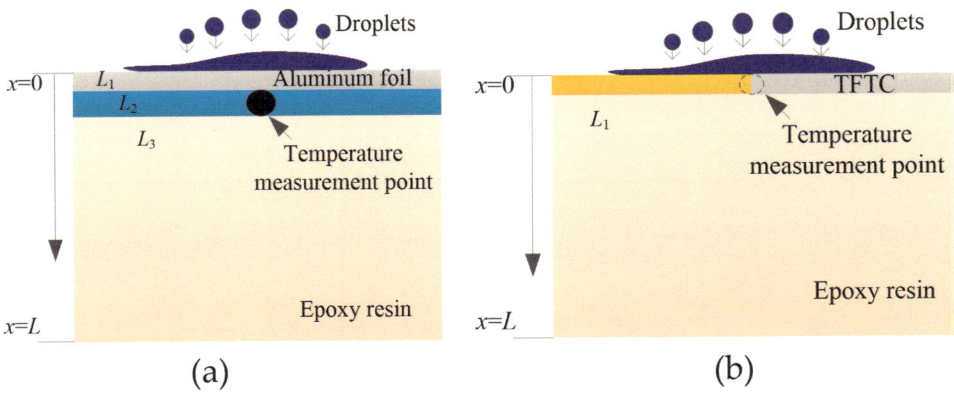

Figure 1.
Schematic of the TFTC and FTC measurements with single temperature sensor [28]: (a) TFTC measurement and (b) FTC measurement.

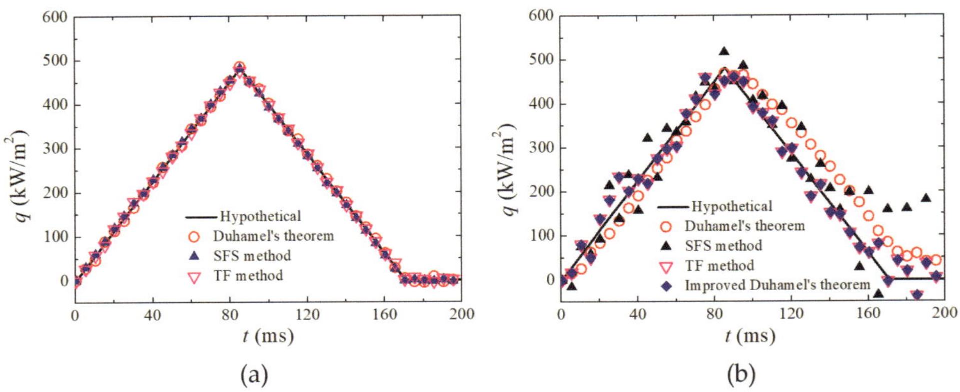

Figure 2.
Results of the comparison between estimated heat fluxes (scatters) and hypothetical ones (solid lines) [28]. (a) single-layer IHCP with single TFTC and (b) three-layer IHCP with single FTC.

substrate (~5 mm), as shown in **Figure 1(b)**. The thermal paste with thickness of 100 μm is placed between the aluminum foil and the epoxy resin, ensuring good thermal contact and providing mechanical support.

A hypothetical triangular pulse heat flux is commonly used to examine the accuracy and sensitivity to random noises of the algorithms [10, 13]. This given heat flux is first set as the surface boundary condition, and then the temperature adding random noise (ε = 1°C) at the measured point corresponding to the two different measurement methods is obtained by solving the direct heat conduction problem. New surface heat flux can be predicted based on the calculated temperature through different algorithms.

Figure 2 shows the new estimated heat fluxes using different algorithms with random noise using TFTC and FTC measurements. The heat fluxes calculated by Duhamel's theorem, SFS, and the transfer function method all agreed well with the hypothetical ones using the TFTC measurement. When using FTC measurement (**Figure 2(b)**), the estimated heat flux obtained from Duhamel's theorem and SFS methods changed nonlinearly over time. Furthermore, they can hardly match the exact heat flux well. Thus, they were unsuitable for estimating the surface heat flux in the indirect surface temperature measurement with FTC. In comparison, both the transfer function and improved Duhamel's theorem can provide linear and accurate results that match the given heat flux exactly.

2.5 1D heat flux estimation

Figure 3 shows the variation in surface temperature as a function of time, measured by the direct measurement method with the TFTC and the indirect measurement method with FTC. The temperature first decreased rapidly when the R404A droplets impinge on the substrate surface, following a relatively slow change as the thin liquid film forms on the surface. It began to resume to the ambient temperature after the liquid film completely evaporated. It is notable that the surface temperature measured by the TFTC measurement decreased faster than that by the indirect FTC measurement method, which showed a definite delay.

Figure 4(a) depicts the time-varied surface heat flux predicted by Duhamel's theorem, SFS method, and transfer function method with TFTC measurement. It can be seen that all the three algorithms predicted very similar results. They first increased rapidly to their maximum values, then dropped quickly to a certain value (~350 kW/m^2), and finally gradually decreased to zero. Note that the estimated result by the SFS method increased faster and had a slightly greater maximum heat flux than that obtained by the other two algorithms.

Figure 4(b) represents the heat flux predicted by different algorithms under the FTC measurement. As mentioned above, Duhamel's theorem was inappropriate for predicting surface heat flux directly from internal measured temperature. Thus, the

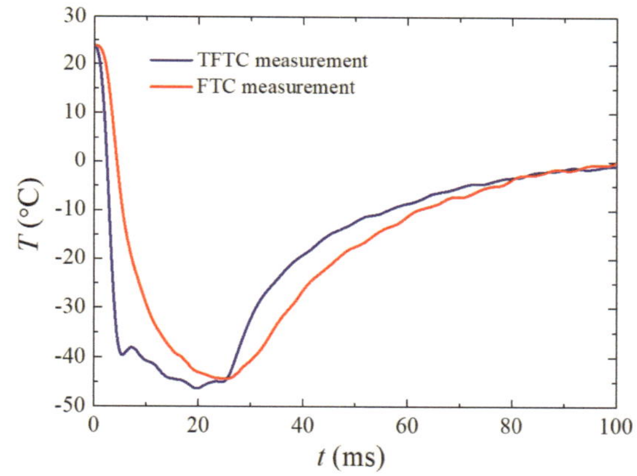

Figure 3.
Variations of surface temperature measured by TFTC and FTC measurements [28].

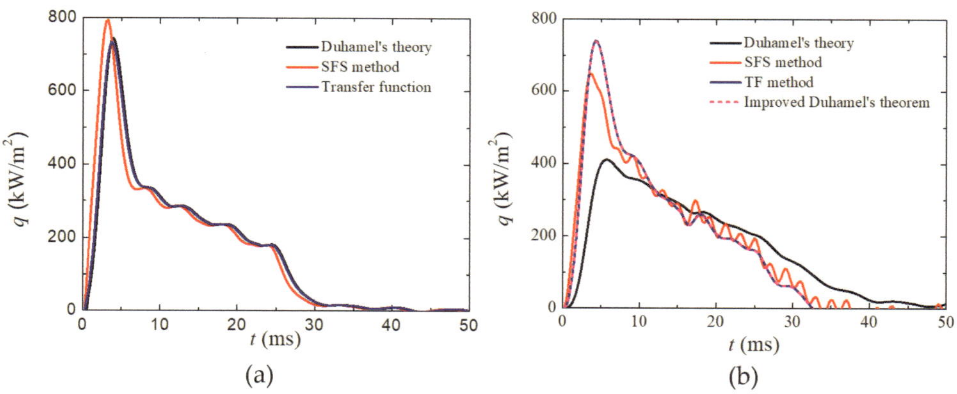

Figure 4.
Variations of surface heat flux predicted by different algorithms [28]. (a) TFTC measurement and (b) FTC measurement.

q_{max} and the magnitude of its increase rate by Duhamel's theorem were both far lower than the case with other algorithms. It seems that the heat flux was significantly delayed and reduced. Other surface heat fluxes predicted by the transfer function and newly improved Duhamel's theory had almost the same varying history, with a larger q_{max} than that predicted by the SFS method.

3. 2D algorithm for heat flux estimation

Although the TF and improved Duhamel's theorem can accurately estimate the surface heat flux with multilayer mediums, it still failed to solve 2D IHCPs. Moreover, large nonuniformity radial and temporal surface temperature variations during CSC often occurred [14, 32]. Therefore, lateral heat transfer must be considered. The 2D algorithm for 2D multilayer medium is urgent to be developed. In this chapter, the 2D filter solution was proposed and validated based on the three-layer FTC temperature measurement.

3.1 2D filter solution and optimal comparison criterion

The 2D general model with K heat fluxes, J temperature sensors, and I layers is shown in **Figure 5**. The multiple surface heat fluxes are assumed to be uniform in a specific upper lateral surface area (e.g., $0 \leq x \leq x_1$). The temperature sensors are placed at the lateral midpoint of each uniform surface heat flux range in the ith layer, and the interval between two thermocouples is Δx. All sensors are placed at an identical depth ($y = H_T$) from the upper surface. The adiabatic boundary condition is considered on the three other sides ($x = 0$, $x = W$, and $y = H$). The initial temperature of the substrates is T_0.

The discrete solution of temperature for direct heat conduction problems with K heat fluxes and J sensors can be presented in a matrix form as follows [37].

$$\mathrm{T} = \mathbf{X}\mathrm{q} \tag{16}$$

where **T** is the real temperature matrix, **X** is the sensitivity matrix, and **q** denotes multiple heat fluxes. The sensitivity matrix **X** is [19]

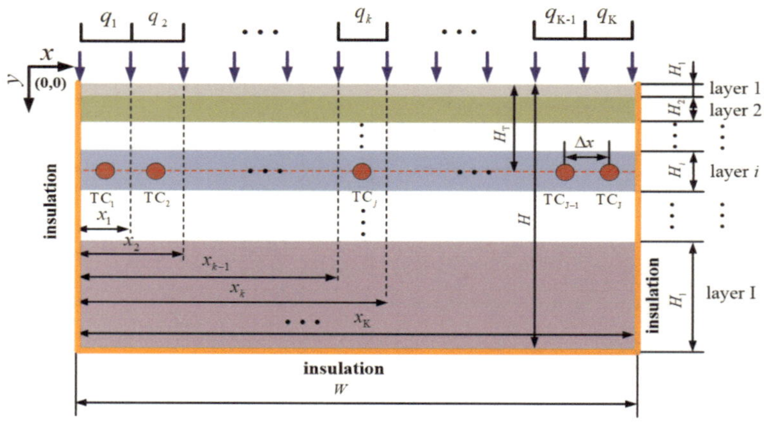

Figure 5.
Geometry model of a 2D multilayer IHCP [36].

$$\mathbf{X} = \begin{bmatrix} \mathrm{a}(1) & 0 & 0 & \cdots & 0 \\ \mathrm{a}(2) & \mathrm{a}(1) & 0 & \cdots & 0 \\ \mathrm{a}(3) & \mathrm{a}(2) & \mathrm{a}(1) & \cdots & \vdots \\ \vdots & \vdots & \vdots & \ddots & 0 \\ \mathrm{a}(\mathrm{N}) & \mathrm{a}(\mathrm{N}-1) & \mathrm{a}(\mathrm{N}-2) & \cdots & \mathrm{a}(1) \end{bmatrix} \tag{17}$$

where

$$\mathrm{a}(n) = \begin{bmatrix} a_{11}(n) & a_{12}(n) & a_{13}(n) & \cdots & a_{1\mathrm{K}}(n) \\ a_{21}(n) & a_{22}(n) & a_{23}(n) & \cdots & a_{2\mathrm{K}}(n) \\ a_{31}(n) & a_{32}(n) & a_{33}(n) & \cdots & a_{3\mathrm{K}}(n) \\ \vdots & \vdots & \vdots & \ddots & \vdots \\ a_{\mathrm{J}1}(n) & a_{\mathrm{J}2}(n) & a_{\mathrm{J}3}(n) & \cdots & a_{\mathrm{JK}}(n) \end{bmatrix} \tag{18}$$

$$a_{jk}(n) = \frac{\partial\left(T\left(x_{\mathrm{TC}_j}, \mathrm{W_T}, t\right) - T_0\right)}{\partial\left[q_k(1)\right]} = \varphi_{jk}(n) \tag{19}$$

where N and n are total time steps and current time step. Excessive temperature ϕ can be presented as follows:

$$\phi_{jk}(n) = T\left(x_{\mathrm{TC}_j}, H_\mathrm{T}, nt_\mathrm{d}\right)\Big|_{q_k=1,\ q_1=q_2=\cdots\cdots=q_\mathrm{K}=0} - T_0 \tag{20}$$

where t_d is the time step. The sum of the squares of the errors between the estimated and measured temperatures plus one-order regularization, S, is

$$\mathrm{S} = (\mathrm{Y\text{-}T})'(\mathrm{Y\text{-}T}) + \alpha_\mathrm{t}[\mathrm{H_t q}]'[\mathrm{H_t q}] + \alpha_\mathrm{s}[\mathrm{H_s q}]'[\mathrm{H_s q}] \tag{21}$$

where **Y** is the measured temperature matrix. α_t and α_s are regularization parameters with respect to temporal and spatial terms. The superscript ‘denotes the transpose of a matrix. $\mathbf{H_t}$ and $\mathbf{H_s}$ are temporal and spatial regularization matrixes. Minimizing S, estimated heat flux matrix $\hat{\mathbf{q}}$ using the entire domain data can be obtained [8]:

$$\hat{\mathrm{q}} = [\mathrm{X'X} + \alpha_\mathrm{t}\mathrm{H_t}'\mathrm{H_t} + \alpha_\mathrm{s}\mathrm{H_s}'\mathrm{H_s}]^{-1}\mathrm{X'Y} = \mathrm{FY} \tag{22}$$

Generally, the 2D multilayer IHCPs are solved layer by layer, starting from the ith layer with the known experimental measured temperature. The heat flux is estimated at the interface with the $(i-1)$th layer. Thereafter, the interface heat flux between the $(i-1)$th and $(i-2)$th layers can be calculated by using the known interface heat flux as the input. Finally, the surface heat flux is estimated [18]. The interface heat fluxes $\hat{\mathbf{q}}_i$ between the ith and $(i-1)$th layers can be described as

$$\hat{\mathrm{q}}_\mathrm{i} = [\mathrm{X_i}'\mathrm{X_i} + \alpha_\mathrm{t}\mathrm{H_t}'\mathrm{H_t} + \alpha_\mathrm{s}\mathrm{H_s}'\mathrm{H_s}]^{-1}\mathrm{X_i}'\mathrm{Y} = \mathrm{F_i Y} \tag{23}$$

Notably, $\mathbf{X}_i$ is different from **X** in the solution of single-layer IHCP. $\mathbf{X}_i$ is calculated by using (I – i + 1) layers below the ith layer. The estimated interface temperature $\hat{\mathbf{T}}_i$ yields

$$\widehat{\mathrm{T}}_i = \mathbf{X}_i\hat{\mathrm{q}}_i \tag{24}$$

The interface heat fluxes ($\hat{\mathbf{q}}_i$) and temperature ($\hat{\mathbf{T}}_i$) are used as the known boundary condition to solve the solution for the $(i-1)$th layer, where the heat fluxes are unknown at $y = H_1 + H_2 + \cdots + H_{i-2}$. The interface heat fluxes $\hat{\mathbf{q}}_{i-1}$ between the $(i-1)$th and $(i-2)$th layers can be expressed as

$$\hat{\mathrm{q}}_{i-1} = \left[\mathbf{X}_{i,\,\mathrm{d}}{}'\mathbf{X}_{i,\,\mathrm{d}} + \alpha_t\mathbf{H}_\mathbf{t}{}'\mathbf{H}_\mathbf{t} + \alpha_s\mathbf{H}_\mathbf{s}{}'\mathbf{H}_\mathbf{s}\right]^{-1}\mathbf{X}_\mathrm{d}{}'\widehat{\mathrm{T}}_i - \mathbf{X}_{i,\,\mathrm{u}}\hat{\mathrm{q}}_i = \mathbf{F}_{i,\,\mathrm{d}}\widehat{\mathrm{T}}_i - \mathbf{X}_{i,\,\mathrm{u}}\hat{\mathrm{q}}_i \quad (25)$$

where $\mathbf{X}_{i,\mathrm{d}}$ and $\mathbf{X}_{i,\mathrm{u}}$ have a similar form as the sensitivity matrix and $\mathbf{X}$ is for single-layer IHCP. However, excessive temperature ϕ in $\mathbf{X}$ is different. For $\mathbf{X}_{i,\mathrm{d}}$, ϕ yields

$$\phi_{jk}(n) = T\left(x_{\mathrm{TC}_j}, H_1 + H_2 + \cdots + H_{i-1}, nt_\mathrm{d}\right)\big|_{q_k=1,\; q_1=q_2=\cdots\cdots=q_K=0} - T_0 \quad (26)$$

For $\mathbf{X}_{i,\mathrm{u}}$, ϕ can be described as

$$\phi_{jk}(n) = T\left(x_{\mathrm{TC}_j}, H_1 + H_2 + \cdots + H_{i-1}, nt_\mathrm{d}\right)\big|_{q_k=1,\; q_1=q_2=\cdots\cdots=q_K=0} - T_0 \quad (27)$$

By substituting Eq. (23) into Eq. (25), the estimated interface heat flux is derived as

$$\hat{\mathrm{q}}_{i-1} = \mathbf{F}_{i,\,\mathrm{d}}\left(\mathbf{X}_i - \mathbf{X}_{i,\,\mathrm{u}}\right)\mathbf{F}_i\mathbf{Y} = \mathbf{F}_{i-1}\mathbf{Y} \quad (28)$$

Interface heat flux $\hat{\mathbf{q}}_{i-2}, \hat{\mathbf{q}}_{i-3}, \cdots \hat{\mathbf{q}}_2$ can be calculated by repeating Eqs. (25)-(28). Thereafter, surface heat flux matrix $\hat{\mathbf{q}}$ can be estimated. The solutions for 2D single-layer and multilayer IHCPs are then obtained.

Given that most filter coefficients can be disregarded except for those of the $(\mathrm{m_p} + \mathrm{m_f} + 1)$ time step, the solutions for IHCPs (Eqs. (22) and (28)) can be simplified into a general filter solution, thereby allowing the real-time heat flux monitoring and small computational load [8, 19, 38]:

$$q_k(n) = \sum_{m=1}^{(\mathrm{m_p}+\mathrm{m_f}+1)\times \mathrm{J}} \left(f_{k,\,m} Y_{(n-\mathrm{m_p}-1)\times \mathrm{J}+m}\right) \quad (29)$$

where $f_{k,m}$ denotes the filter coefficient in the mth column from one row of $\mathbf{F}$ associated with the kth unknown heat flux. Notably, α_t and α_s significantly affect the accuracy of the estimated heat flux, which should be determined for a specific IHCP [9, 18, 19]. These parameters can be optimized by the optimal comparison criterion developed for solving 2D single-layer and multilayer IHCPs [36]. The sum of the squares of the errors, $\mathrm{R_q}$, between the estimated and real heat fluxes was used to examine the accuracy of the calculation:

$$\mathrm{R_q} = (\hat{\mathrm{q}} - \mathbf{q})'(\hat{\mathrm{q}} - \mathbf{q}) = (\mathbf{FY} - \mathbf{q})'(\mathbf{FY} - \mathbf{q}) \quad (30)$$

A more efficient means is to minimize the expected value of $\mathrm{R_q}$ [19, 38] as follows:

$$E\left(\mathrm{R_q}\right) = [(\mathrm{FX} - \mathrm{I})\mathrm{q}]'[(\mathrm{FX} - \mathrm{I})\mathrm{q}] + \sigma_Y^2 tr(\mathbf{F}'\mathbf{F}) \quad (31)$$

where E is the expected value, tr denotes the sum of the diagonal elements, and σ_Y (0.01°C [38]) is the uniform measurement errors. $E(\mathrm{R_q})$ contains two components: $E_{q,bias}$:

$$E_{q,\,bias} = [(\mathrm{FX} - \mathrm{I})\mathrm{q}]'[(\mathrm{FX} - \mathrm{I})\mathrm{q}] \quad (32)$$

and $E_{q,rand}$ in the filter form

$$E_{q,\ rand} = \sigma_Y^2 tr\left(\mathbf{F}^T\mathbf{F}\right) = \sigma_Y^2\left(\mathrm{N} - \left(\mathrm{m_p} + \mathrm{m_f} + 1\right)\right)\sum_{k=1}^{\mathrm{K}}\sum_{j=1}^{\left(\mathrm{m_p}+\mathrm{m_f}+1\right)} f_{k,j\text{-}1*\mathrm{J}+k}^2 \tag{33}$$

To examine the accuracy of the estimated heat fluxes, the mean relative error (MRE) between estimated and hypothetical heat fluxes was employed as follows:

$$q_{MRE} = \frac{1}{\mathrm{N}-\mathrm{m_f}}\sum_{n=1}^{\mathrm{N}-\mathrm{m_f}}\left|\frac{\hat{q}(n) - q(n)}{q(n)}\right| \tag{34}$$

3.2 Validation of 2D filter solution

The 2D three-layer geometry (**Figure 6**) with six FTC sensors was used to validate the accuracy of surface heat flux estimated by 2D filter solution. For FTC measurement, the multiple surface heat fluxes were assumed to be uniform in a specific upper lateral surface area (e.g., $0 \le x \le x_1$) and different from each other. Temperature was measured from the spray center (x = 0 mm, TC_1) to the periphery (x = 10 mm, TC_6), and the lateral distance between two FTCs was Δx = 2 mm. The geometry width with TFTC measurement was W = 11 mm, which is half that of the spray diameter because the computational domain is symmetric.

The optimal comparison criterion developed for 2D IHCPs was employed in this study to optimize the Tikhonov regularization parameters (α_t and α_s). As shown in **Figure 7**, logarithmic random component $E_{q,rand}$ increased as the regularization parameters α_t and α_s decreased. $E_{q,bias}$ and $E(\mathrm{R_q})$ reached their minimum value at $\alpha_t = \alpha_s = 10^{-9}$. Eventually, the optimum regularization parameters were determined to be $\alpha_t = \alpha_s = 10^{-9}$.

As shown in **Figure 8**, the estimated heat fluxes agreed well with the hypothetical ones. However, small deviations were observed at the descent stage after the maximum heat flux for q_1 to q_6. The q_{MRE} of q_1–q_6 for TFTC measurement was 2.63%, 2.66%, 2.76%, 2.78%, 3.00%, and 5.18%, respectively. Adding the random noise at the measured point, the maximum q_{MRE} increased to 3.71%, which indicated that the accuracy and stability of the filter solution are satisfactory.

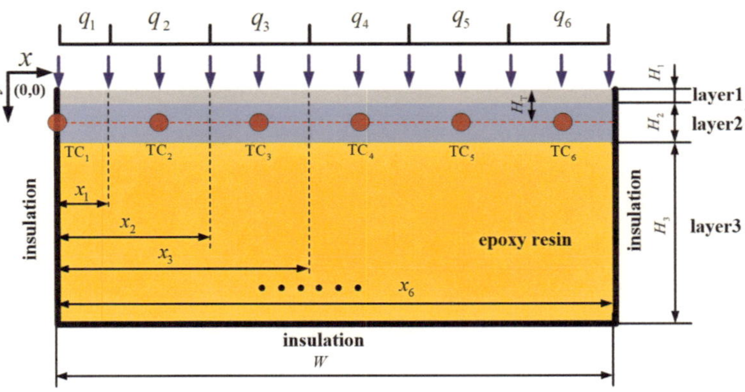

Figure 6.
Geometry model of three-layer IHCP with six FTCs [36].

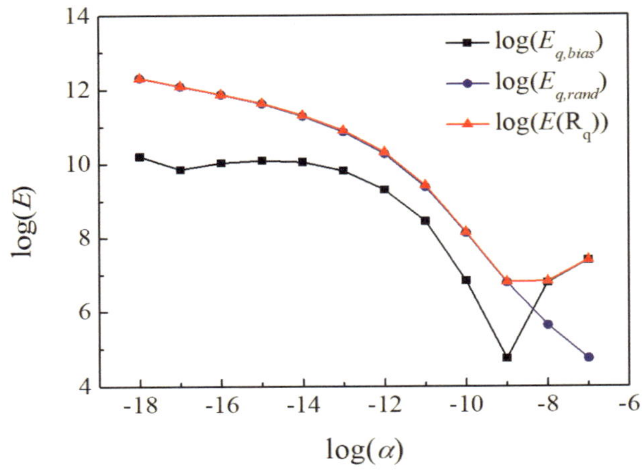

Figure 7.
Geometry model of three-layer IHCP with six FTCs [36].

3.3 Comparison between 2D filter solution and 1D improved Duhamel's theorem

To investigate the importance considering lateral heat transfer, the estimated surface heat flux (**Figure 9(a)**) and simulated temperature (**Figure 9(b)**) using the estimated surface heat flux as boundary, calculated by 2D filter solution and 1D improved Duhamel's theorem, were compared taking the example of 2D three-layer geometry with FTC measurement. For FTC measurement containing aluminum film with high heat conductivity coefficient ($\lambda = 236\ \text{W}\bullet\text{m}^{-1}\bullet\text{K}^{-1}$), the maximum heat flux calculated by the 1D method is underestimated by 60% than that calculated by 2D filter solution, indicating that the lateral heat transfer cannot be disregarded, especially when the heat conductivity coefficient of the material is large. As shown in **Figure 9(b)**, the simulated temperature using estimated surface heat flux as boundary calculated by 2D filter solution agreed well with measured temperature, indirectly

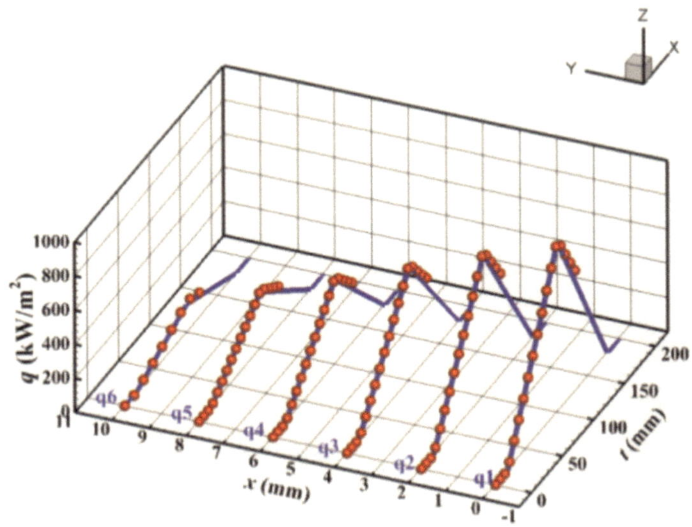

Figure 8.
Results of the comparison between estimated heat fluxes (scatters) and hypothetical ones (solid lines) for three-layer IHCP with FTCs [36].

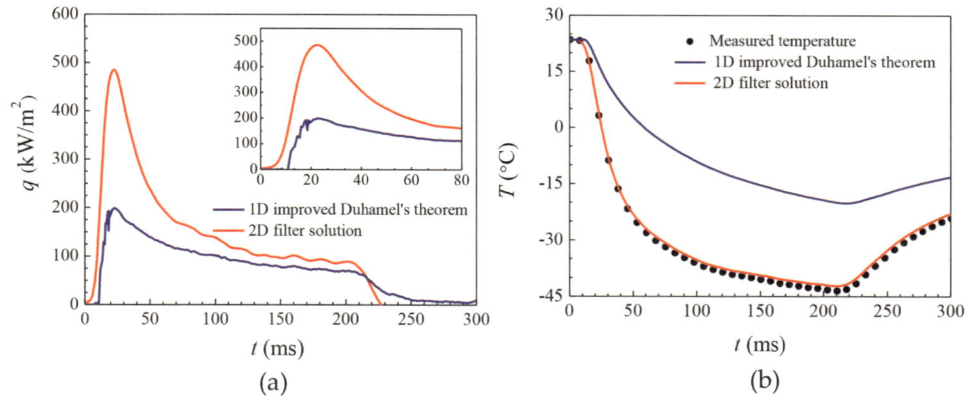

Figure 9.
Comparative results at spray center calculated by 2D filter solution and 1D improved Duhamel's theorem [36]. (a) surface heat flux and (b) temperature at sensor location.

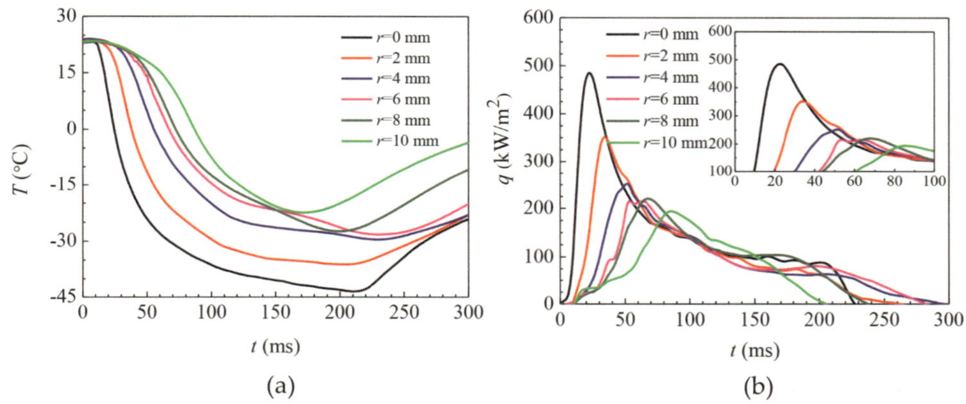

Figure 10.
Variations of surface temperature and heat flux with FTC measurement [36]. (a) measured surface temperature and (b) estimated surface heat flux.

indicating the accuracy of 2D filter solution. However, the large simulated temperature deviation was observed using 1D improved Duhamel's theorem, owing to the inaccurate estimated surface heat flux disregarding lateral heat transfer.

The dynamic internal temperature measured by six FTCs with Δt = 50 ms and L = 30 mm is depicted in **Figure 10(a)**. The temperature histories were similar, but differences existed in the specific values. The minimum temperatures (T_{min}) at r = 0, 2, 4, 6, 8, and 10 mm were −43.42, −36.13, −29.55, −28.22, −27.32, and −22.45°C, respectively. Additionally, the measured temperature was lower at the spray center (r = 0 mm) than periphery. **Figure 10(b)** presents the estimated heat fluxes calculated by the filter solution for 2D three-layer IHCP with FTC measurement. The estimated heat flux profiles at different locations in this figure were also similar. However, a large difference was observed in heat fluxes at different lateral locations. The best cooling capacity was found at the spray center (r = 0–2 mm).

4. Clinical potential of spray cooling by low GWP R1234yf

The extremely high global warming potential (GWP = 1430) of commercially used cryogen R134a with boiling point of −26.1°C will cause severe environmental

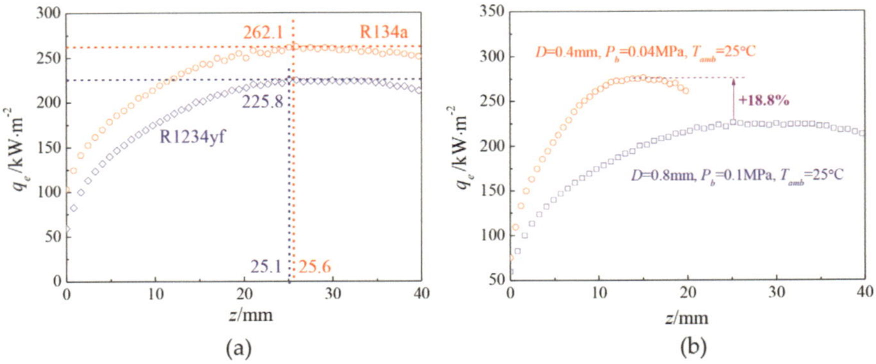

Figure 11.
Variations of effective surface heat flux as a function of spray distance [39]. (a) effective surface heat flux with R134a and R1234yf and (b) effective surface heat flux before and after cooling enhancement with R1234yf.

hazards, and the substitution of R134a in clinical application is urgent. R1234yf with boiling point of −29.5°C may be a potential candidate for environment protection due to its low GWP (<1). In this chapter, the clinical potential of R1234yf substitution for R134a was investigated.

Using the maximum surface heat flux correlation obtained by experimental spray characteristics (droplet temperature, velocity, and diameter) and surface heat transfer performance (surface heat flux calculated by 2D filter solution), **Figure 11(a)** shows the variations of effective surface heat flux with different R134a and R1234yf as a function of spray distance. The effective surface heat flux (q_e) was obtained by multiplying the maximum surface heat flux, and cooling concentration within the radius of 2 mm is the interested area [39]. The effective surface heat flux by using R134a and R1234yf increased firstly due to the droplet temperature reduction as spray distance increased. Then, q_e reaches their maximum value and finally decreases slowly. The maximum q_e of R134a and R1234yf is 262.1 and 225.8 kW/m^2 at the optimal spray distances of 25.6 and 25.1 mm (see **Figure 11(a)**), respectively. The substitution of R1234yf for R134a can produce remarkable reduction of global warming potential to <1. However, the cooling capacity should be enhanced for the clinical application in laser treatment, owing to the 13.8% reduction in effective heat flux (from 262.1 to 225.8 kW/m^2). Therefore, the enhancement of cooling capacity is necessary for the implementation of R1234yf in clinical laser treatment of PWS.

According to our experience, two simple ways are available to enhance the cooling capacity, i.e., changing the nozzle diameter and decreasing the back pressure by decreasing the boiling point of cryogens [40, 41]. As shown in **Figure 11(b)**, the enhancement of effective surface heat flux at different spray distances was remarkable. After reducing the nozzle diameter and decreasing the back pressure, q_e increases rapidly due to violent evaporation, and the peak value of the effective heat flux (268.3 kW/m^2) is increased by 18.8%. This result is comparable with that of R134a under 1 atm, which proves the potential of R1234yf in the clinic CSC for the laser treatment of PWS.

5. Conclusions

Several algorithms including the SFS, TF, and Duhamel's theorem methods were analyzed and compared in predicting time-varying surface heat flux during CSC.

Duhamel's theorem was improved to get the accurate results through the transformation of internal temperature into surface temperature, when the indirect surface temperature measurement (FTC) method is used. A hypothetical triangular pulse heat flux was employed to analyze the accuracy and sensitivity to noise of the algorithms under TFTC and FTC measurements. The estimated result of Duhamel's theorem and SFS method widely deviated from the given heat flux under the three-layer FTC measurement method, whereas the transfer function and improved Duhamel's theorem all provided the exact estimated heat flux.

The 2D filter solution was proposed to solve a general 2D multilayer IHCP for the estimation of surface heat flux. An optimal comparison criterion was employed to optimize the key parameters, namely, α_t and α_s. Six hypothetical triangular heat fluxes and random temperature errors of 1°C were employed to analyze the accuracy and sensitivity of the filter solution for 2D three-layer IHCPs with FTC measurement. The q_{MRE} values for FTC measurement with and without the random temperature errors were all within the acceptable range, which validates the good accuracy and stability of the filter solutions. The maximum heat flux calculated by the 1D method was underestimated by 60% than that calculated by 2D filter solution considering lateral heat transfer. The 2D filter solution was more accurate than the 1D method. Moreover, lateral heat transfer should not be ignored, especially when the heat conductivity coefficient of the material is large.

The surface heat transfer characteristics of spray cooling with R134a and R1234yf were investigated based on 2D filter solution. The maximum effective heat fluxes $q_{e,\max}$ were 262.1 and 225.8 kW/m^2 for R134a and R1234yf at different spray distances of 25.6 and 25.1 mm. Through the cooling enhancement of reducing the nozzle diameter and decreasing the back pressure, $q_{e,\max}$ of R1234yf was increased by 18.8% (D = 0.4 mm and P_b = 0.04 MPa). The enhanced $q_{e,\max}$ is a bit higher than that of R134a in normal condition, which provides a theoretical basis for potential application of low GWP R1234yf in clinics.

Acknowledgements

This work was supported by the National Natural Science Foundation of China (51727811) and Fundamental Research Funds for the Central Universities.

Conflict of interest

The authors declare no conflict of interest.

Author details

Bin Chen*, Jia-Meng Tian and Zhi-Fu Zhou
State Key Laboratory of Multiphase Flow in Power Engineering, Xi'an Jiaotong University, Xi'an, China

*Address all correspondence to: chenbin@xjtu.edu.cn

References

[1] Ko WL, Quinn RD, Gong L. Finite-element reentry heat-transfer analysis of space shuttle Orbiter. Nasa Sti/recon Technical Report N, 87; 1987

[2] Calame JP. High heat flux thermal management of microfabricated upper millimeter-wave vacuum electronic devices. In: Vacuum Electronics Conference, 2008. IVEC 2008; IEEE International; 2008. pp. 50-51

[3] Wikström P, Blasiak W, Berntsson F. Estimation of the transient surface temperature and heat flux of a steel slab using an inverse method. Applied Thermal Engineering. 2007;**27**: 2463-2472

[4] Taler J, Duda P, Węglowski B, Zima W, Grądziel S, Sobota T, et al. Identification of local heat flux to membrane water-walls in steam boilers. Fuel. 2009;**88**:305-311

[5] Lee C-H, Hourng L-W, Lin K-W. Mathematical model predicting the critical heat flux of nuclear reactors. Journal of Computer Science. 2012;**8**:1996

[6] Tian JM, Chen B, Li D, Zhou ZF. Transient spray cooling: Similarity of dynamic heat flux for different cryogens, nozzles and substrates. International Journal of Heat and Mass Transfer. 2017;**108**:561-571

[7] Sumita I, Olson P. Rotating thermal convection experiments in a hemispherical shell with heterogeneous boundary heat flux: Implications for the Earth's core. Journal of Geophysical Research. 2002;**107**:ETG 5-1-ETG 5-18

[8] Najafi H, Woodbury KA, Beck JV. Real time solution for inverse heat conduction problems in a two-dimensional plate with multiple heat fluxes at the surface. International Journal of Heat and Mass Transfer. 2015; **91**:1148-1156

[9] Beck JV. Filter solutions for the nonlinear inverse heat conduction problem. Inverse Problems in Science and Engineering. 2008;**16**:3-20

[10] Beck JV, Blackwell B, Clair CRS Jr. Inverse Heat Conduction: Ill-Posed Problems. New York: James Beck; 1985

[11] Hansen PC. Regularization tools: A Matlab package for analysis and solution of discrete ill-posed problems. Numerical Algorithms. 1994;**6**:1-35

[12] Alifanov O. Solution of an inverse problem of heat conduction by iteration methods. Journal of Engineering Physics and Thermophysics. 1974;**26**:471-476

[13] Fernandes AP, dos Santos MB, Guimarães G. An analytical transfer function method to solve inverse heat conduction problems. Applied Mathematical Modelling. 2015;**39**:6897-6914

[14] Franco W, Liu J, Wang GX, Nelson JS, Aguilar G. Radial and temporal variations in surface heat transfer during cryogen spray cooling. Physics in Medicine and Biology. 2005; **50**:387-397

[15] Bianchi D, Donatelli M, Serracapizzano S. Optimal order conditions for filter based regularization methods, with applications to variants of the Tikhonov method. Mathematics. 2014;**6**:59-71

[16] Cabeza JMG, Rodríguez AC. Filter digital form of two future temperature methods for the inverse heat conduction: A spectral comparison. International Journal for Numerical Methods in Biomedical Engineering. 2010;**26**:554-573

[17] Peng YM, Wang KL, Zhang HC. Improved Tikhonov regularization

research and applied in inverse problem. Applied Mechanics and Materials. 2011; **50-51**:447-450

[18] Najafi H, Woodbury KA, Beck JV. A filter based solution for inverse heat conduction problems in multi-layer mediums. International Journal of Heat and Mass Transfer. 2015;**83**:710-720

[19] Woodbury KA, Beck JV. Estimation metrics and optimal regularization in a Tikhonov digital filter for the inverse heat conduction problem. International Journal of Heat and Mass Transfer. 2013;**62**:31-39

[20] Najafi H, Woodbury KA, Beck JV, Keltner NR. Real-time heat flux measurement using directional flame thermometer. Applied Thermal Engineering. 2015;**86**:229-237

[21] Hsieh S-S, Luo S-Y. Droplet impact dynamics and transient heat transfer of a micro spray system for power electronics devices. International Journal of Heat and Mass Transfer. 2016;**92**:190-205

[22] Dou R, Wen Z, Zhou G. Heat transfer characteristics of water spray impinging on high temperature stainless steel plate with finite thickness. International Journal of Heat and Mass Transfer. 2015;**90**:376-387

[23] Bhattacharya P, Samanta A, Chakraborty S. Spray evaporative cooling to achieve ultra fast cooling in runout table. International Journal of Thermal Sciences. 2009;**48**:1741-1747

[24] Chen S, Liu J, Liu X, Hou Y. An experimental comparison of heat transfer characteristic between R134-a and R22 in spray cooling. Experimental Thermal and Fluid Science. 2015;**66**: 206-212

[25] Nelson JS, Milner TE, Anvari B, Tanenbaum BS, Kimel S, Svaasand LO, et al. Dynamic epidermal cooling during pulsed laser treatment of port-wine stain. A new methodology with preliminary clinical evaluation. Archives of Dermatology. 1995;**131**:695-700

[26] Zhou Z, Chen B, Wang Y, Guo L, Wang G. An experimental study on pulsed spray cooling with refrigerant R-404a in laser surgery. Applied Thermal Engineering. 2012;**39**:29-36

[27] Li D, Chen B, Wu WJ, Wang GX, He YL. Multi-scale modeling of tissue freezing during cryogen spray cooling with R134a, R407c and R404a. Applied Thermal Engineering. 2014;**73**: 1489-1500

[28] Zhou ZF, Xu TY, Chen B. Algorithms for the estimation of transient surface heat flux during ultra-fast surface cooling. International Journal of Heat and Mass Transfer. 2016;**100**:1-10

[29] Aguilar G, Wang GX, Nelson JS. Effect of spurt duration on the heat transfer dynamics during cryogen spray cooling. Physics in Medicine and Biology. 2003;**48**:2169-2181

[30] Zhou Z, Wu W, Chen B, Wang G, Guo L. An experimental study on the spray and thermal characteristics of R134a two-phase flashing spray. International Journal of Heat and Mass Transfer. 2012;**55**:223

[31] Zhou Z-F, Chen B, Wang R, Wang G-X. Comparative investigation on the spray characteristics and heat transfer dynamics of pulsed spray cooling with volatile cryogens. Experimental Thermal and Fluid Science. 2017;**82**:189-197

[32] Wang R, Zhou Z, Chen B, Bai F, Wang G. Surface heat transfer characteristics of R404A pulsed spray cooling with an expansion-chambered nozzle for laser dermatology. International Journal of

Refrigeration-Review International Froid. 2015;**60**:206-216

[33] Beck J, Blackwell B, Haji-Sheikh A. Comparison of some inverse heat conduction methods using experimental data. International Journal of Heat and Mass Transfer. 1996;**39**:3649-3657

[34] Tunnell JW, Torres JH, Anvari B. Methodology for estimation of time-dependent surface heat flux due to cryogen spray cooling. Annals of Biomedical Engineering. 2002;**30**:19-33

[35] Hahn DW, Ozisik MN. Heat Conduction. New Jersey: John Wiley & Sons; 2012

[36] Tian J-m, Chen B, Zhou Z-f. Methodology of surface heat flux estimation for 2D multi-layer mediums. International Journal of Heat and Mass Transfer. 2017;**114**:675-687

[37] Beck JV, Blackwell B, Clair CRS. Inverse Heat Conduction: Ill-Posed Problems. New York: Wiley; 1985

[38] Beck JV, Woodbury KA. Inverse heat conduction problem: Sensitivity coefficient insights, filter coefficients, and intrinsic verification. International Journal of Heat and Mass Transfer. 2016;**97**:578-588

[39] Chen B, Tian J, Wang R, Zhou Z. Theoretical study of cryogen spray cooling with R134a, R404A and R1234yf: Comparison and clinical potential application. Applied Thermal Engineering. 2019;**148**:1058-1067

[40] Zhou Z-f, Chen B, Wang R, Bai F-l, Wang G-x. Coupling effect of hypobaric pressure and spray distance on heat transfer dynamics of R134a pulsed flashing spray cooling. Experimental Thermal and Fluid Science. 2016;**70**:96-104

[41] Aguilar G, Franco W, Jie LM, Svaasand LO, Nelson JS. Effects of hypobaric pressure on human skin: Implications for cryogen spray cooling (part II). Lasers in Surgery and Medicine. 2005;**36**:130-135

Chapter 8

Equation of State

Eman Mohamed Mansour

Abstract

An equation of state (EOS) is a thermodynamic expression that relates pressure (P), temperature (T), and volume (V). This equation is used to describe the state of reservoir fluids at given conditions. The cubic equations of state (CEOS) such as Van der Waals, Redlich-Kwong, Soave, and Peng-Robinson are simple models that have been widely used in the oil industry. This chapter expressed literature for EOS that varies from simple expressions to multiple constant and convoluted types of equations. Many attempts have been made to describe the thermodynamic behavior of fluids to predict their physical properties at given conditions. So, several forms of the equation of state have been presented to the oil industry in order to calculate reservoir fluid properties. The heat exchanger is important in wildly fields as in aerospace, petrochemical industry, refrigeration, and other fields. The optimization design of the heat exchanger is a great significance to industry process to reduce production cost, realize energy conservation, and reduce energy consumption.

Keywords: EOS types, EOS importance, EOS history, cubic equation of state, heat transfer

1. Introduction

In the last six decades, petroleum engineers realized the importance of using EOS for PVT modeling in addition to the following [1]:

1. Predicting physical properties of the black oil in the laboratories is very expensive.

2. Difficulty to obtain a representative sample due to reservoir nature or operation issues.

3. Insufficient sample volume in order to obtain a complete analysis.

4. Error in lab analyses.

5. Samples are nonrepresentative (not monophasic sample, OBM contamination, etc.).

6. Lab report quality check.

7. Estimating the hydrocarbon reserves.

1.1 Classification of equation of state

There are different types of EOS which fall into three categories:

1.1.1 First class of EOS

These equations are basically cubic equation of state. The cubic equations of state such as the Van der Waals [2], Redlicha and Kwong [3], Soave-Redlich-Kwong [4], and Peng-Robinson [5] equations give reasonable results for the thermodynamic behavior of real fluids.

1.1.2 Second class of EOS

These EOS are non-cubic in form. They are providing accurate results for both vapor and liquid phases. The Benedict et al. [6] equation is a good example for this class equation.

1.1.3 Third class of EOS

These are nonanalytical EOS that are highly constrained for some specific fluids [7]. Even though they are constrained, they are capable of expressing real fluid thermodynamic properties precisely.

Among all these EOS, the first-class EOS is more useful because it provides an analytical solution than the more complex and complicated non-cubic second type and nonanalytical third type that require time-consuming iterative calculations. In general, the overall performance in fluid properties prediction is somewhat better using the Soave-Redlich-Kwong (SRK) equation than using the Redlich-Kwong (RK) and Van der Waals EOS [8].

2. Development history of the equation of state

Several forms of EOS have been presented to the petroleum industry to estimate hydrocarbon reservoir fluid properties and sought to a better representation of the PVT relationship for fluids [9].

In 1662, Robert Boyle (Boyle's law) discovered that for a constant temperature, there is an inverse relationship between volume of gas and its pressure ($P \propto V^{-1}$). In 1780, Jacques Charles (Charles's Law) showed that the volume of gas is proportional to the absolute temperature at a constant pressure ($V \propto T$). In 1834, Clapeyron combined these two results into the ideal gas law, PV = RT [10], assuming that the molecules are very far and have no attractive or repulsive forces between them and elastic collisions between these molecules. This equation is known as the ideal gas law and/or the general gas law. It is expressed mathematically as [11].

$$PV = nRT \tag{1}$$

where P: absolute pressure, psia; V: volume, ft^3; T: absolute temperature, °R; R: the universal gas constant (10.73159 ft^3 psia $°R^{-1}$ lb-mole^{-1}; n: number of moles of gas, lb-mole.

For gases at low pressures, the ideal gas law is a convenient satisfactory tool. The application of the ideal gas law at higher pressures may lead to errors up to 500%, compared to 2–3% at atmospheric pressure. Real gases behave differently than ideal gases, the reason for this deviation is that the ideal gas law was derived under the assumption that the volume of molecules is very small and neither molecular attraction nor repulsion exists between them, and this is not the real case. In order to write an equation of state for a real gas, a correction factor has to be inserted into the ideal gas equation [12]:

$$P\,V = Zn\,R\,T \tag{2}$$

where Z: the correction factor which is known as the compressibility factor.

The equation has different names, such as the compressibility equation and/or the real gas equation [13]. A review of recent advances in the empirical cubic EOS field is presented next [11]. Van der Waals [2] is one of the earliest attempts to represent the behavior of real gases by an equation, where the two assumptions were made for the ideal gas EOS:

1. The gas molecule volume is very small compared to the volume of the container.

2. There are no attractive or repulsive forces between the gas molecules or the walls of the container.

Van der Waals attempted to eliminate these assumptions in the development of an empirical EOS for the real gases.

First assumption elimination: the gas molecules occupy a considerable fraction of the volume at higher pressures, and the volume of the molecules (b) is subtracted from the actual molar volume (V) to give the following expression:

$$p = \frac{RT}{v - b} \tag{3}$$

Second assumption elimination: he added corrective term (a), denoted by (a/V^2), in order to account for the attractive forces between molecules.

Van der Waals introduced the following equation (Eq. (4)):

$$\left(p + \frac{a}{{V_M}^2}\right)(V_M - b) = RT \tag{4}$$

where a: attraction parameter; b: repulsion parameter.

The symbol "a" is considered a measure of the intermolecular attractive forces between the molecules. "b" is known as the co-volume and considered to reflect the volume of molecules [2]. The "a" and "b" values can be obtained from the critical properties of the fluid [14], where the repulsion pressure, $p_{repulsion}$, is represented by the term $RT/(V_m - b)$, and the attraction pressure, $p_{attraction}$, is described by $a/{V_m}^2$. The Van der Waals equation of state despite its simplicity, while it is provide a correct description and qualitative of the PVT substances behavior in the liquid and gaseous phases. Yet, it is not accurate enough to be suitable for design purposes. The equation of state approach for calculating physical properties and phase equilibrium proved to be a powerful tool, and much energy was devoted to the development of new and accurate equations of state [11]. Other researchers began attempts to improve Van der Waals equation of state for over 100 years. Usually a change of the molecular attraction term (a/V_mM^2) was proposed. Clausius in 1880 [15] proposed that the molecular attraction term was inversely proportional to temperature [16]:

$$\left[p + \frac{a}{T(V_M + c)^2}\right](V_M - b) = RT \tag{5}$$

The addition of a fourth constant (c) enabled better agreement with data. However, mathematical manipulations required in thermodynamic calculations were more difficult. So Berthelot in 1899 [17] removed the constant (c), resulting in the following equation:

$$\left(p + \frac{a}{TV_M{}^2}\right)(V_M - b) = RT \tag{6}$$

Dieterici in 1899 [18] handled the temperature dependence of the molecular attraction term in a different manner [6]:

$$\left[P\ EXP\left(\frac{a}{V_M RT}\right)\right](V_M - b) = RT \tag{7}$$

Lorentz in 1881 [19] addressed the molecular volume term [20]:

$$\left(p + \frac{a}{V_{M^2}}\right)\left(V_M - \frac{bV_M}{V_M + b}\right) = RT. \tag{8}$$

Wohl in 1927 [21] considered the effect of temperature on the molecular attraction term:

$$\left(P + \frac{a}{TV_M(V_M - b)} - \frac{c}{T^2 V_M{}^3}\right)(VM - b) = RT \tag{9}$$

The constants a, b, and c in the equations above have different values for different substances. Several investigations proposed virial-type of EOS. Kammerlingh-Onnes in 1901 [22] proposed the virial equation of state as follows [23]:

$$PV_M = RT\left[1 + \frac{B}{V_M} + \frac{C}{V_M{}^2} + \ldots\ldots\right] \tag{10}$$

where B and C are not constants which are functions of temperature and are called the second and third virial coefficients. Beattie and Bridgeman in 1927 published a five-constant equation that gives a satisfactory representation of volumetric properties except in the critical region [24]:

$$P = \frac{RT}{V_{M2}}\left(1 - \frac{c}{V_M T^3}\right)\left[V_M + B_o\left(1 - \frac{b}{V_M}\right)\right] - \frac{A_{o(1-a/V_M)}}{V_M} \tag{11}$$

Benedict et al. [6] suggested a multiparameter equation of state known as the Benedict-Webb-Rubin (BWR) equation [6]:

$$P = \frac{RT}{V_M} + \frac{B_o RT - A_o - C_o/T^2}{V_M{}^2} + \frac{bRT - a}{V_M{}^3} + \frac{a\alpha}{V_M{}^6} + \frac{c}{T^2 V_M{}^3}\left(1 + \frac{\gamma}{V_M{}^2}\right)EXP\left(\frac{-\gamma}{V_M{}^2}\right) \tag{12}$$

This equation may be considered a modification of the Beattie-Bridgeman equation of state where A_O, B_O, C_O, a, b, c, α, and γ are eight adjustable parameters.

The BWR equation could treat critical components and was able to work in the critical area. However, the BWR equation suffers from some disadvantages [25]. Perhaps, the most important model for the modification of the Van der Waals equation of state is the Redlich-Kwong (RK) (1949) which is demonstrated by an adjustment of the Van der Waals's attraction term (a/V_m^2) and includes the system temperature explicitly. They could improve the prediction of the physical and volumetric properties of the vapor phase. In RK EOS, the attraction pressure term was replaced with a generalized temperature-dependent term (Eq. (13)) [3]:

$$\rho = \frac{RT}{V - b} - \frac{\alpha}{V(V + b)\sqrt{T}} \tag{13}$$

For pure substances, the equation parameters *a* and *b* are usually expressed as.

$$b = \Omega b\, R\, T_c / P_c \tag{14}$$

$$a = \Omega\, \alpha\, R^2 T_c^{\,2.5} / P_c \tag{15}$$

where Ωa = 0.42747 and Ωb = 0.08664.
Replacing the molar volume (V) in Eq. (13) with (ZRT/P) and rearranging give.

$$Z^3 - Z^2 + \left(A - B - B^2\right) Z - AB = 0 \tag{16}$$

where

$$B = \frac{bp}{RT} \tag{17}$$

$$A = \frac{ap}{R^2 T^{2.5}} \tag{18}$$

Three real roots in the two-phase region are yielded. The largest root corresponds to the compressibility factor of the gas phase, Z_v, while the smallest positive root corresponded to that of the liquid, Z_L [11].

For mixtures, the equation parameters a and b are usually expressed as a_m and b_m for a hydrocarbon liquid mixture with a composition of x_i:

$$a_m = \left[\sum_{i=1}^{n} Xi\, \sqrt{a_i}\right]^2 \tag{19}$$

$$b_m = \sum_{i=1}^{n} [X_i b_i] \tag{20}$$

a_m and b_m for a hydrocarbon gas mixture with a composition of y_i:

$$a_m = \left[\sum_{i=1}^{n} y_i\, \sqrt{a_i}\right]^2 \tag{21}$$

$$b_m = \sum_{i=1}^{n} [y_i b_i] \tag{22}$$

where n: number of components in the mixture; a_i: Redlich-Kwong a parameter for the i component; b_i: Redlich-Kwong b parameter for the i component; b_m:

parameter b for mixture; x_i: mole fraction of component i in the liquid phase; y_i: mole fraction of component i in the gas phase.

Replacing the molar volume (V) in Eq. (13) with (ZRT/P) and rearranging give.

$$Z^3 - Z^2 + \left(A - B - B^2\right) Z - AB = 0 \tag{23}$$

where

$$B = \frac{b_m p}{RT} \tag{24}$$

$$A = \frac{a_m P}{R^2 T^{2.5}} \tag{25}$$

Then the compressibility factor of the gas phase or the liquid can be calculated.

Joffe and Zudkevitch [26] showed that a substantial improvement in the representation of fugacity of gas mixtures could be obtained by treating interaction parameters as empirical parameters [26]. Spear et al. [27] also states that the RK equation of state could be used to calculate the vapor-liquid critical properties of binary mixtures [28]. Chueh and Prausnitz [29] showed that the RK equation can be adapted to predict both vapor and liquid properties. Spear et al. [28] gave seven examples of systems for which the vapor-liquid critical properties of hydrocarbon mixtures could be calculated by using the RK equation of state. Carnahan and Starling [30] used the Redlich-Kwong equation of state to calculate the gas-phase enthalpies for a variety of substances [30]. Their results showed that the Redlich-Kwong equation was a significant improvement over the Van der Waals equation. Other workers applied the Redlich-Kwong equation to the critical properties and the high-pressure phase equilibria of binary mixtures. The results showed that the accuracy of the Redlich-Kwong equation of state calculations for ternary systems was only slightly less than that for the constituent binaries [31].

The success of the Redlich-Kwong equation has been the impetus for many further empirical improvements. One of the milestones in developing of CEOS was reported by Soave [4]. His development in the evaluation of the parameter in the attraction pressure term for the RK equation is shown in (Eq. (22)). Soave replaced the term $(a/T^{0.5})$ in Eq. (22) with a more general temperature-dependent term, denoted by a α (T), to give

$$\rho = \frac{RT}{V - b} - \frac{a\,\alpha(T)}{V(V + b)} \tag{26}$$

where α(T) is a dimensionless factor. Soave used vapor pressures of pure components to introduce an expression for the temperature correction parameter α(T). At temperatures other than the critical temperature, the correction parameter α(T) was defined by the following equation:

$$\alpha(T) = \left[1 + m\left(1 - \sqrt{T_r}\right)\right]^2 \tag{27}$$

Soave correlated the parameter “m” with the centric factor (ω) to give.

$$m = 0.480 + 1.574\varpi, \; - 0.176\varpi^2 \tag{28}$$

where T_r: reduced temperature, °R; ω: a centric factor of the substance; T: system temperature, °R.

For pure substances the equation parameters a and b are usually expressed as.

$$b = \Omega b\ R\ T_c/P_c \tag{29}$$

$$a = \Omega a\ R^2 T_c{}^2/P_c \tag{30}$$

In general, most EOS inputs are only the critical properties, and a centric factor of each components is shown in **Table 1**.

where Ωa and Ωb are the SRK dimensionless pure component parameters:

Ωa = 0.42747.

Ωb = 0.08664.

Replacing the molar volume (V) in the equation with (ZRT/p) and rearranging give the compressibility factor Z:

$$Z^3 - Z^2 + \left(A - B - B^2\right) Z - AB = 0 \tag{31}$$

where

$$B = \frac{b_m p}{\mathrm{RT}} \tag{32}$$

$$A = \frac{a_m P}{(\mathrm{RT})^2} \tag{33}$$

For mixtures, the equation parameters a and b are usually expressed as a_m and b_m for a hydrocarbon liquid mixture with a composition of x_i:

$$a_m = \sum_i \sum_j \left[x_i x_j \sqrt{a_i a_j \alpha_i \alpha_j}\left(1 - k_{ij}\right)\right. \tag{34}$$

$$b_m = \sum_i [Xi\ bi] \tag{35}$$

The following is the calculation for a_m and b_m for a gas mixture with a composition of y_i:

$$a_m = \sum_i \sum_j \left[y_i y_j \sqrt{a_i a_j \alpha_i \alpha_j}\left(1 - k_{ij}\right)\right. \tag{36}$$

$$b_m = \sum_i [yi\ bi] \tag{37}$$

A binary interaction parameter (BI), classically noted as k_{ij}, is usually involved in the “a” parameter expression to provide more flexibility to the EOS and designed to characterize any binary system formed by components i and j in the hydrocarbon mixture [32]. Vidal and Daubert [33], Graboski and Daubert [34], and Slot-Petersen [35] suggested that no BIs were required for hydrocarbon systems. However, with no hydrocarbons present, binary interaction parameters can improve the phase in volumetric behavior predictions of the mixture by the SRK EOS for compressibility factor calculations of the gas or the liquid phases [34, 36, 37]. The equilibrium ratio, K_i, that is, K_i = yi /xi, can be redefined in terms of the fugacity of component:

Compound	Formula	Molecular weight	Critical pressure (psla)	Critical temperature (°F)	Critical volume (ft^3/lb)	Liquid specific gravity (water = 1)	Gas specific (air = 1)	Acentric factor
Methane	CH_4	16.042	667.0	−116.66	0.0985	(0.3)	0.55400	0.0115
Ethane	C_2H_6	30.069	706.6	89.92	0.0775	0.35643	1.03830	0.0994
Propane	C_3H_8	44.096	615.5	205.92	0.0728	0.50738	1.52270	0.1529
Isobutane	C_4H_{10}	58.122	527.9	274.41	0.0715	0.56295	2.00710	0.1865
n-Butane	C_4H_{10}	58.122	550.9	305.55	0.0703	0.58408	2.00710	0.2003
Isopentane	C_5H_{12}	72.149	490.4	369	0.0685	0.62460	2.49140	0.2284
n-Pentane	C_5H_{12}	72.149	488.8	385.8	0.0676	0.63113	2.49140	0.2515
n-Hexane	C_6H_{14}	86.175	436.9	453.3	0.0688	0.66404	2.97580	0.2993
n-Heptane	C_7H_{16}	100.202	396.8	512.9	0.0682	0.68819	3.46020	0.3483
n-Octane	C_8H_{18}	114.229	360.7	564.2	0.0673	0.70698	3.94450	0.3977
n-Nonane	C_9H_{20}	128.255	330.7	610.8	0.693	0.72186	4.42890	0.4421
n-Decane	$C_{10}H_{22}$	142.282	304.6	652.2	0.0703	0.73406	4.91330	0.4875
Carbon monoxide	CO	28.01	506.7	−220.63	0.0527	0.79265	0.96720	0.0510
Carbon dioxide	CO_2	44.01	1070.0	87.76	0.0343	0.82203	1.51970	0.2239
Hydrogen sulfide	H_2S	34.082	1306.5	212.81	0.0462	0.80269	1.17690	0.1010
Air	—	28.9586	551.9	−220.97	0.0458	0.87603	1.00000	—
Hydrogen	H_2	2.0159	190.7	−399.9	0.5319	0.07087	0.06961	−0.2140
Oxygen	O_2	31.9988	731.4	−181.43	0.0367	1.14230	1.10500	0.0222
Nitrogen	N_2	28.0135	492.5	−232.53	0.0511	0.80687	0.96740	0.0372
Water	H_2O	18.0153	3200.1	705.1	0.04975	1.00000	0.62210	0.3443

Table 1.
Physical properties of each components.

$$K_i = \frac{\left[f_i^L/(XiP)\right]}{\left[f_i^V/(y_iP)\right]} = \frac{\Phi_i^L}{\Phi_i^v} \tag{38}$$

where f_i^v = fugacity of component "i" in the gas phase; f_i^L = fugacity of component "i" in the liquid phase; Φ_i^v = fugacity coefficient of component "i" in the vapor phase; Φ_i^L = fugacity coefficient of component "i" in the liquid phase.

Soave proposed the following expression for the fugacity coefficient of component i in the liquid phase:

$$ln\left(\frac{f_i^L}{X_iP}\right) = ln\left(\Phi_i^L\right) = \frac{b_i(Z^l-1)}{b_m}\ ln\ \left(Z^L-B\right) - \left(\frac{A}{P}\right)\left[\frac{2\psi_i}{a_m} - \frac{b_i}{b_m}\right]\ ln\ \left[1+\frac{B}{Z^L}\right] \tag{39}$$

where

$$\psi_j = \sum_j \left[x_j\sqrt{a_ia_j\alpha_i\alpha_j}\left(1-k_{ij}\right)\right] \tag{40}$$

$$a_m = \sum_i\sum_j \left[x_ix_j\sqrt{a_ia_j\alpha_i\alpha_j}\left(1-k_{ij}\right)\right. \tag{41}$$

Fugacity coefficient of component i in the gas phase:

$$\ln\left(\phi_j\right) = \frac{b_i\left(Z^i-1\right)}{b_m} - \ln\left(Z^i-B\right) - \left(\frac{A}{B}\right)\left[\frac{2\psi_i}{a_m} - \frac{b_i}{b_m}\right]\ln\left[1+\frac{B}{Z^i}\right] \tag{42}$$

where:

$$\psi_j = \sum_j \left[y_j\sqrt{a_ia_j\alpha_i\alpha_j}\left(1-k_{ij}\right)\right] \tag{43}$$

$$a_m = \sum_i\sum_j \left[y_iy_j\sqrt{a_ia_j\alpha_i\alpha_j}\left(1-k_{ij}\right)\right. \tag{44}$$

3. Heat exchanger

Heat exchanger is an energy (heat) exchange equipment, where it transfers the heat from a working medium to another working medium. Knowing heat exchanger is important in wildly fields as in aerospace, petrochemical industry, refrigeration, and other fields. The optimization design of the heat exchanger is a great significance to industry process to reduce production cost, realize energy conservation, and reduce energy consumption [38]. The development technique for different types of the heat exchanger has been reviewed by many researchers. The development method can be by two ways: passive method and active method. The passive method is to generate swirling flow and disturb the thermal boundary layer by installing vortex generator or tabulators such as baffle, rib, winglet, wing, etc.

The active method is to add the external power to increase efficiency and heat transfer rate such as vibration. So the use of the active method must consider both benefit of the system and additional power cost [39].

3.1 Optimization techniques of heat exchanger

The optimization techniques of heat exchanger can be shown at three different stages as the following [40]:

1. Identification of the lowest initial cost of a heat exchanger design that meets the process specifications.

2. Identification of a heat exchanger design that will work most acceptably over the plant lifetime.

3. Identification of the minimum total cost of the process by choosing heat exchangers system and auxiliary components that will make the best plant process specifications.

3.2 Criteria of the ideal heat exchanger

The dissimilarities of optimization techniques levels can be understood if we list the required criteria of the ideal heat exchanger as follows [41]:

1. The heat exchangers should be flexible enough to meet specifications process such as normal fouling transients and seasonal and diurnal changes in service stream temperatures.

2. Special requirements as weight, length, or inventory standards are important for heat exchangers especially in retrofit applications.

3. The heat exchanger must endure operation under standard and foreseeable operating conditions, maintain the mechanical stresses of manufacturing transport, and minimize the effects of fouling and corrosion.

4. The heat exchanger must be maintainable, cleaning, repair or replacement and its components as gaskets and tubes with minimum downtime.

5. The exchanger must achieve process specifications, i.e., achieve any changes in the thermal conditions by allowable pressure drops.

6. There are other requirements, such as experience, capability of operating, maintenance personnel, and manufacturing time.

7. The exchanger should cost as little as possible.

4. Fluid flow and heat transfer mechanism

The heat transfer and fluid flow mechanism in porous media is important in engineering and industrial fields such as petroleum and chemical engineering [42]. This mechanism occurs for many studies such as in microchannels and nanofluids.

Example 1: In case of study characteristics of fluid flow and heat transfer in the (100) silicon microchannel heat sink, the heat convection capabilities in the phase changes as well as in a single-phase flow and the mechanism of bubble nucleation. In the heat transfer characteristics, the results illustrate that changing in the phase process in the microchannels reduces environment working temperature and absorbs the heat. Six different microchannel geometries are selected for the heat transfer experiment as shown in **Table 2**.

Figure 1 shows that the decreasing wall temperature phenomenon during the phase change is the same as Peng and Wang [43].

On the aspect of fluid flow characteristics, the effects of the viscosity and friction coefficient of the fluid in the microchannels are much significant than the macros. Where the specifications of the sink are registered in **Table 2**, Chip 1–4 are prepared for fluid flow experiment. The friction factor is decreasing with the power of Reynolds number as shown in **Figure 2** [44].

1. Nanoparticle types and concentration in the base fluids

2. Changing in thermophysical properties such as viscosity, surface tension, thermal conductivity, density, and heat capacity

3. The operation condition especially the mass and heat fluxes

4. The modification of wettability and capillary wicking force surface roughness

During the last 2 years, there were some review papers which outlined the subject of boiling heat transfer using nanofluids as a new category in thermal fluids.

Chip name	Width (μm) W_c	Depth (μm) H_c	Hydraulic diameter (μm)	Number of channels
Chip 1	400	260	221	10
Chip 2	300	130	150	13
Chip 3	250	184	134	15
Chip 4	200	148	109	19
Chip 5	150	113	83	25
Chip 6	100	78	57	38

Table 2.
Specification of the sink.

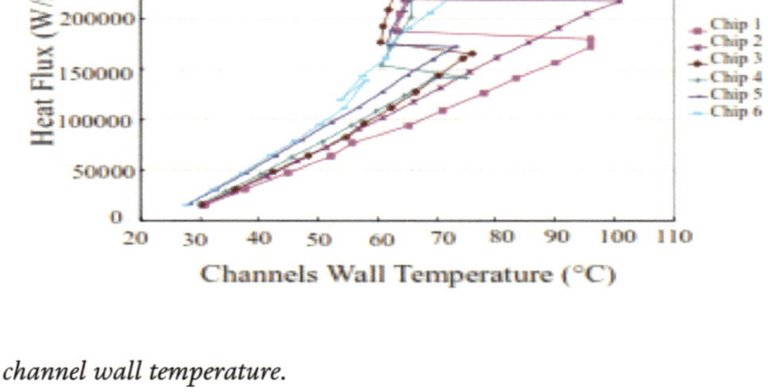

Figure 1.
The heat flux and channel wall temperature.

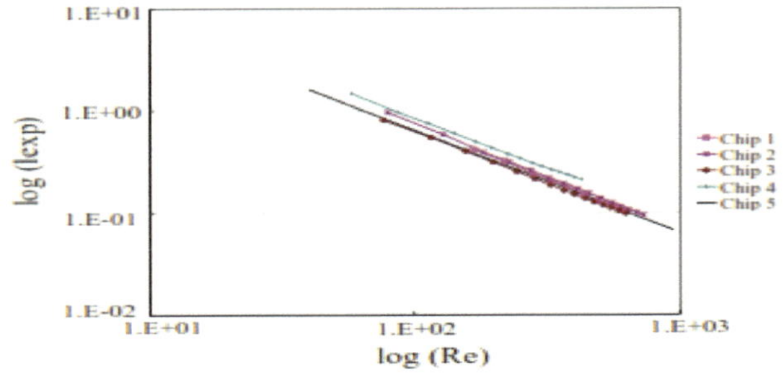

Figure 2.
Exponential relations between the friction factor and the Reynolds number. Example 1: in the case of study characteristics of fluid flow and heat transfer, nanofluid is widely utilized in numerous industrial applications such as boiler tubes, evaporators, and cooling of reactors in a nuclear power plant. The main parameters that directly influence on the heat transfer performance are listed as follows [45].

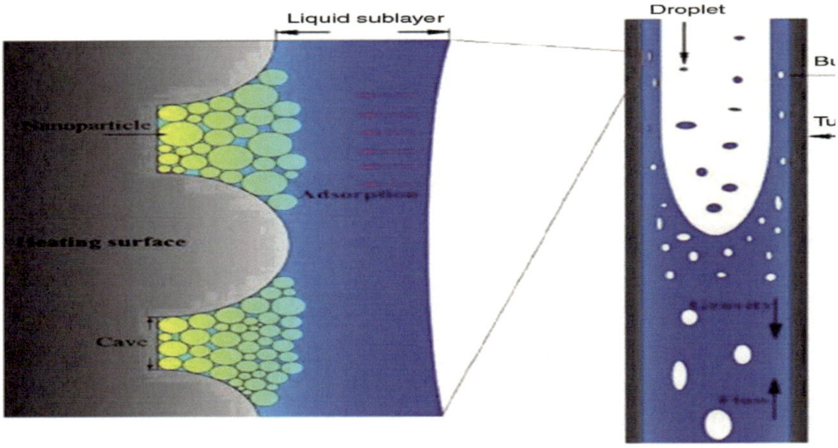

Figure 3.
Illustration of the mechanism of flow boiling CHF using nanofluid introduced.

Available results reported that the effect of nanoparticles on the flow boiling HTC is conflicting, but the CHF could enhance on 50%. During the boiling process, parameters such as flow pattern and pressure drop were affected by the deposition of nanoparticles [46]. Authors concluded that using nanofluids might intensify the boiling HT and CHF, depending on many parameters related to additives, nanoparticles, geometry concentration, and fluid properties [47].

Their work shows how the nanofluids can achieve high heat flux with small temperature differences throughout the boiling process, which modify the critical heat flux [48]. All article reviewers said that nanofluids is a complicated phenomenon and it is not fully understood from mechanism of boiling heat transfer and two-phase flow. Collected studies show enhancement in CHF, and its improvement could achieve more than 50% [49]. HTC behavior could increase or decrease during flow boiling and pool, and it depends on several parameters related to surface texture on thermophysical properties [50]. Wang et al. shows the mechanism of depositing nanoparticles on the heating surface and creating the porous layer as shown in **Figure 3**. The results show that the CHF of flow boiling is enhanced up to 18% as compared to conventional fluid. This enhancement increases with increasing some parameters, for example, the pressure system and the channel diameter [45].

Example 2: in the case of study fluid flow and heat transfer characteristics using nanofluid in a single-phase turbulent flow by using helically corrugated tubes, pitch-to-diameter ratio (P/DH = 0.18, 0.22 and 0.27) and rib-height-to-diameter ratio (e/DH = 0.02, 0.04 and 0.06) of helically corrugated tubes on the heat transfer enhancement, isothermal friction, and thermal performance factor in a concentric tube heat exchanger are examined. Results illustrate that the thermal performance of the corrugated tube and heat transfer are increased as compared to those of the smooth tube. The rate increase in heat transfer rate is between 123 and 232%, depending on the rib height/pitch ratios and Reynolds number. The friction factor (average) of the corrugated tube is between 1.46 and 1.93 over the smooth tube [51].

Abbreviations

EOS	equation of state
BWR	Benedict-Webb-Rubin
RK	Redlich-Kwong
SRK	Soave-Redlich-Kwong
CEOS	cubic equations of state
MSRKE	modified Soave-Redlich-Kwong equation of state
p	system pressure, psia
P_c	critical pressure, psia
P_r	reduced pressure, psia
p_i	initial pressure, psia
T	system temperature, °F
T_c	critical temperature, °R
T_r	reduced temperature, °R
T_i	initial temperature, °F
a	equation of state attraction parameter
b	equation of state repulsion parameter
B	second virial coefficient
C	third virial coefficient
a, b, c, A_0, B_0, C_0	constant in Benedict-Webb-Rubin equation
a, b, c, A_0, B_0	constant in Beattie and Bridgeman equation of state
$f^v{}_i$	fugacity of component i in the gas phase
$f^L{}_j$	fugacity of component j in the liquid phase
$Ø^v{}_i$	fugacity coefficient of component i in the vapor phase
$Ø^L{}_i$	fugacity coefficient of component i in the liquid phase
x	mole fraction of gas phase
y	mole fraction of liquid phase
k	equilibrium ratio for the component
K_{ij}	interaction coefficient parameter
A, B	parameter in Soave-Redlich-Kwong equation of state
a_T	temperature-dependent coefficient in Soave-Redlich-Kwong equation of state
a_c	constant coefficient in Soave-Redlich-Kwong equation of state
a_{Ti}	temperature-dependent coefficient of component i
m	parameter in Soave-Redlich-Kwong equation of state
Z	compressibility factor
Z_i	the mole fraction of component in the mixture
Z_L	compressibility factor of the liquid phase
Z_v	compressibility factor of the gas phase
ω	a centric factor of the substance
n_l	number of moles in liquid phase
n_v	number of moles in gas phase

n_t	total number of moles of mixture
F_l	one mole liquid of total fraction
F_v	one mole vapor of total fraction
X_i	mole fraction in the liquid phase
y_i	mole fraction in the vapor phase
γ	constant in Benedict-Webb-Rubin equation of state

Author details

Eman Mohamed Mansour
PVT Lab, Production Department, Egyptian Petroleum Research Institute, Cairo, Egypt

*Address all correspondence to: emanmansour84@yahoo.com

References

[1] Asadisaghandi J, Tahmasebi P. Comparative evaluation of back-propagation neural network learning algorithms and empirical correlations for prediction of oil PVT properties in Iran oilfields. Journal of Petroleum Science and Engineering. 2011;**78**(2):464-475

[2] Van der Waals JD. Over de Continuiteit van den Gas-en Vloeistoftoestand. Leiden: A.W. Sijthoff; 1873

[3] Redlich O, Kwong JN. On the thermodynamics of solutions. V. An equation of state. Fugacities of gaseous solutions. Chemical Reviews. 1949;**44**(1):233-244

[4] Soave G. Equilibrium constants from a modified RedlichKwong equation of state. Journal of Chemical Engineering Science. 1972;**27**:1197-1203

[5] Peng DY, Robinson DB. A new two-constant equation of state. Journal of Engineering Chemical Fundamental. 1976;**15**:59-64

[6] Benedict M, Webb GB, Rubin LC. An empirical equation for thermodynamic properties of light hydrocarbons and their mixtures I. Methane, ethane, propane and n-butane. The Journal of Chemical Physics. 1940;**8**(4):334-345

[7] Goodwin RD. Provisional Thermodynamic Functions of Propane, from 85 to 700 K at Pressures to 700 Bar. National STANDARD Reference Data System. Institute for Basic Standards. 1977

[8] Patel NC, Teja AS. A new cubic equation of state for fluids and fluid mixtures. Chemical Engineering Science. 1982;**37**(3):463-473

[9] Whitson CH, Brulé MR. Phase Behavior, Henry L. Doherty Memorial Fund of AIME, Society of Petroleum Engineers; 2000

[10] Atllhan M. A new cubic equation of state [M.Sc. thesis]. Texas: A&M University; 2004

[11] Ahmed T. Equations of State and PVT Analysis. Gulf Professional Publishing; 2013

[12] Mansour E, Desouky S, Batanoni M, Mahmoud M, Farag A, El-Dars F. Modification proposed for SRK equation of state. Oil and Gas Journal. 2012;**110**(6):78-91

[13] Yao Y, Xie T, Gao Y. Physical Chemistry Handbook. Shanghai Science and Technology Publishing Company; 1985. p. 706

[14] Wei YS, Sadus RJ. Equations of state for the calculation of fluid phase equilibria. AIChE Journal. 2000;**46**:169-196

[15] Clausius R. Ueber des verhalten der kohlensaure in begug auf druck, volumen and temperatur. Annual Physical Chemistry. 1880;**9**:337

[16] William D. The Properties of Petroleum Fluids Handbook. Vol. 148. Tulsa: PennWell Publishing Company; 1994

[17] Berthelot DJ. Sur une méthode purement physique pour la détermination des poids moléculaires des gaz et des poids atomiques de leurs éléments. Journal of Physics. 1899;**8**:263

[18] Dieterici C. Ueber den kritischen Zustand. Annual Physical Chemistry. 1899;**30**:685

[19] Lorentz. Wied. Annual Physical Chemistry. 1881;**12**(127):660

[20] Boyd MF. Studies on the bionomics of north American AnopheLines: Physical and chemical factors in their relation to the distribution of larvae in northeastern North Carolina. American Journal of Hygiene. 1929;**9**(2):6

[21] Wohl KZ. Physical Chemistry. 1927; **133**:305

[22] Kammerlingh-Onnes HK. Commununications from the Physics Laboratory. Vol. 71. Leiden, Holland; 1901

[23] Adamson AW, Gast AP. Physical Chemistry of Surfaces. New York: Interscience; 1967

[24] Beattie JA, Bridgeman OC. A new equation of state for fluids. I. Application to gaseous ethyl ether and carbon dioxide1. Journal of the American Chemical Society. 1927;**49**(7): 1665-1667

[25] Mansour E, Farag A, El-Dars F, Desouky S, Batanoni M, Mahmoud M. Predicting PVT properties of Egyptian crude oils by a modified Soave-Redlich-Kowng equation of state. Egyptian Journal of Petroleum. 2013;**22**(1): 137-148

[26] Joffe J, Zudkevitch D. Fugacity coefficients in gas mixtures containing light hydrocarbons and carbon dioxide. Industrial & Engineering Chemistry Fundamentals. 1966;**5**(4):455-462

[27] Spear RR, Robinson RL, Chao KC. Critical states of mixtures and equations of state. Industrial Engineers Chemical Fundamental. 1969;**8**:2-8

[28] Spear RR, Robinson RL Jr, Chao K-C. Critical states of ternary mixtures and equations of state. Industrial & Engineering Chemistry Fundamentals. 1971;**10**(4):588-592

[29] Chueh PL, Prausnitz JM. Vapour-liquid equilibria at high pressures: Calculation of partial molar volumes in non-polar liquid mixtures. AIChE Journal. 1967;**13**:1099

[30] Carnahan NF, Starling KE. Intermolecular repulsions and the equation of state for fluids. AICHE Journal. 1972;**18**(6):1184-1189

[31] Mansour E, Al-Sabagh A, Desouky S, Zawawy F, Ramzi M. Experimental approach of minimum miscibility pressure for CO_2 miscible flooding: Application to Egyptian oil fields. International Journal of New Technology and Research. 2016;**2**(5):8

[32] Mansour E, Al-Sabagh A, Desouky S, Zawawy F, Ramzi M. A laboratory investigation of carbon dioxide-enhanced oil recovery by focusing on CO_2-oil physical properties. Egyptian Journal of Petroleum. 2019;**28**(1):21-26

[33] Vidal J, Daubert T. Equations of state-reworking the old forms. Journal of Chemical Engineering Society. 1978; **33**:787-791

[34] Graboski MS, Daubert TE. A modified soave equation of state for phase equilibrium calculations. 1. Hydrocarbon systems. Industrial & Engineering Chemistry Process Design and Development. 1978;**17**(4):443-448

[35] Slot-Petersen C. A systematic and consistent approach to determine binary interaction coefficients for the Peng-Robinson equation of state. In: 62nd, Annual Technical Conference of the SPE, Dallas, TX; 27-30 September 1987

[36] Slot-Petersen C. A systematic and consistent approach to determine binary interaction coefficients for the Peng-Robinson equation of state (includes associated papers 20308 and 20393). SPE Reservoir Engineering. 1989;**4**(04): 488-494

[37] Vidal J. Equations of state-reworking the old forms. Fluid Phase Equilibria. 1983;**13**:15-33

[38] Zeugmann T, Poupart P, Kennedy J, Jin X, Han J, Saitta L, et al. Particle Swarm Optimization. Boston, MA: Springer US; 2011. pp. 760-766

[39] Boonloi A, Jedsadaratanachai W. Numerical study on flow and heat transfer mechanisms in the heat exchanger channel with V-orifice at various blockage ratios, gap spacing ratios, and flow directions. Modelling and Simulation in Engineering. 2019; **2019**:5

[40] Rao RV, Patel V. Multi-objective optimization of heat exchangers using a modified teaching-learning-based optimization algorithm. Applied Mathematical Modelling. 2013;**37**(3): 1147-1162

[41] Yao J. A review of industrial heat exchange optimization. In: IOP Conference Series: Earth and Environmental Science. IOP Publishing; 2018

[42] Chatti S, Ghabi C, Mhimid A. Fluid flow and heat transfer in porous media and post heated obstacle: Lattice Boltzmann simulation. International Journal of Heat and Technology. 2016; **34**(3):377-385

[43] Peng X, Peterson G, Wang B. Frictional flow characteristics of water flowing through rectangular microchannels. Experimental Heat Transfer An International Journal. 1994; **7**(4):249-264

[44] Chen Y-T, Kang S-W, Tuh W-C, Hsiao T-H. Experimental investigation of fluid flow and heat transfer in microchannels. 淡江理工學刊. 2004;**7** (1):11-16

[45] Kamel MS, Lezsovits F, Hussein AK. Experimental studies of flow boiling heat transfer by using nanofluids. Journal of Thermal Analysis and Calorimetry. 2019;**138**:1-25

[46] Zhang C, Zheng L, Zhang X, Chen G. MHD flow and radiation heat transfer of nanofluids in porous media with variable surface heat flux and chemical reaction. Applied Mathematical Modelling. 2015;**39**(1): 165-181

[47] Fang X, Chen Y, Zhang H, Chen W, Dong A, Wang R. Heat transfer and critical heat flux of nanofluid boiling: A comprehensive review. Renewable and Sustainable Energy Reviews. 2016; **62**:924-940

[48] Kamel MS, Lezsovits F, Hussein AM, Mahian O, Wongwises S. Latest developments in boiling critical heat flux using nanofluids: A concise review. International Communications in Heat and Mass Transfer. 2018;**98**:59-66

[49] Cheng L, Xia G, Li Q, Thome JR. Fundamental issues, technology development, and challenges of boiling heat transfer, critical heat flux, and two-phase flow phenomena with nanofluids. Heat Transfer Engineering. 2018;**3**:1-36

[50] Moreira TA, Moreira DC, Ribatski G. Nanofluids for heat transfer applications: A review. Journal of the Brazilian Society of Mechanical Sciences and Engineering. 2018;**40**(6):303

[51] Pethkool S, Eiamsa-Ard S, Kwankaomeng S, Promvonge P. Turbulent heat transfer enhancement in a heat exchanger using helically corrugated tube. International Communications in Heat and Mass Transfer. 2011;**38**(3):340-347

Chapter 9

Influence of Input Parameters on the Solution of Inverse Heat Conduction Problem

Rakhab C. Mehta

Abstract

A one-dimensional transient heat conduction equation is solved using analytical and numerical methods. An iterative technique is employed which estimates unknown boundary conditions from the measured temperature time history. The focus of the present chapter is to investigate effects of input parameters such as time delay, thermocouple cavity, error in the location of thermocouple position and time- and temperature-dependent thermophysical properties. Inverse heat conduction problem IHCP is solved with and without material conduction. A two-time level implicit finite difference numerical method is used to solve nonlinear heat conduction problem. Effects of uniform, nonuniform and deforming computational grids on the estimated convective heat transfer are investigated in a nozzle of solid rocket motor. A unified heat transfer analysis is presented to obtain wall heat flux and convective heat transfer coefficient in a rocket nozzle. A two-node exact solution technique is applied to estimate aerodynamic heating in a free flight of a sounding rocket. The stability of the solution of the inverse heat conduction problem is sensitive to the spatial and temporal discretization.

Keywords: analytical solution, inverse heat conduction problem, numerical analysis, deforming grid, heat transfer coefficient, heat flux, random search method

1. Introduction

The basic theory of heat and structure of solid body is associated with the internal energy of matter which in the first law of thermodynamics is referred to as the internal energy concerned with the physical state of the material. The first law of thermodynamics defines that the flowing heat energy is conserved in the absence of heat sources and sinks. It is, therefore, important to study the influence of thermocouple lead wires and distortion due to the thermocouple cavity in solution of the inverse heat conduction problem. According to the second law of thermodynamics, the heat will be transferred from one body to another body only when the bodies are at two different temperatures level and the heat will flow from the point of higher to the point of lower temperature.

A direct solution of transient heat conduction equation with prescribed initial and boundary conditions yields temperature distribution inside a slab of finite thickness. The direct solution is mathematically considered as well-posed because the solution exists, unique and continuously depends on input data. The estimation of unknown parameters from the measured temperature history is called as inverse problem of heat conduction. It is mathematically known as an ill-posed problem since the solution now does not continuously depend on the input data. Measurement data error in temperature, thermal lagging, thermocouple's cavity, signal noise, etc. makes stability problem in the estimation of unknown parameters.

Numerical inversion of the integral solution [1], exact solution [2], numerical techniques [3], least-squares method [4], transform methods [5], different series approach [6], variable time-step size [7] have been applied to solve inverse heat conduction problems. Solutions of the ill-posed inverse heat conduction problem have been presented in detail by Beck et al. [8] and Özisik et al. [9]. Tikhonov regularization method [10] has been described for cross-validation criterion for selecting the regularization parameter to obtain a stable approximation to the solution. Kurpisz et al. [11] have presented series with derivatives with temperature to solve inverse thermal problem. Hensel [12] has described space marching numerical methods to solve inverse heat transfer problem. Various mathematical methods and numerical algorithms for solving inverse heat conduction problems are described and compared by Alifanov [13]. Taler and Duda [14] have presented solutions of direct and inverse heat conduction problems.

Inverse heat conduction analysis provides an efficient tool for estimating the thermophysical properties of materials, the boundary conditions, or the initial conditions. Estimation of surface heat flux has been carried out without [15] and with [16] heat conduction and comparison between them shows discrepancies as high as about 27% [17]. Moving window optimization method [18] has been applied to predict the aerodynamic heating in a free-flight of sounding rocket by comparing numerically calculated and measured temperature history. Howard [19] developed a numerical procedure for estimating the heat flux with variable thermal properties using a single embedded thermocouple. Simultaneous identification of the temperature-dependent thermal conductivity and the asymmetry parameter of the Henyey-Greenstein scattering phase function have been shown by Zmywaczyk and Koniorczyk [20].

The conjugate gradient method with adjoint problem for function estimation iterative technique is used to solve IHCP to estimate heat flux and internal wall temperature of the throat section of the rocket nozzle [21]. Heisler [22] have reported supplementary "short-time" temperature-time charts for the center, mid-location and surface of large plates, long cylinders and spheres for the dimensionless time sub-domain. Convective heat transfer coefficient and combustion temperature in a rocket nozzle is determined using transient-temperature response chart [23].

The solution of transient IHCP can be obtained using analytical or numerical schemes in conjunction with measured temperature-time history. The estimation of the unknown parameters can be carried out by employing gradient or non-gradient methods to predict the unknown parameters in a prescribed tolerance limit. The focus of the present work is to investigate the influence of various parameters on the solution of inverse heat conduction problem.

2. Measurement errors

Experimental difficulties [24] are noticed in implanting thermocouples at the surface for temperature measurements. Temperature response delays have been studied to solve IHCP applied to cooled rocket thrust chamber [25]. The temperature measured inside the slab may delay and damp depending on x_m as illustrated in **Figure 1**. A thermocouple indicates temperature lag behind the actual temperature. The effect of the thermocouple sensor dynamics on prediction of a triangular heat

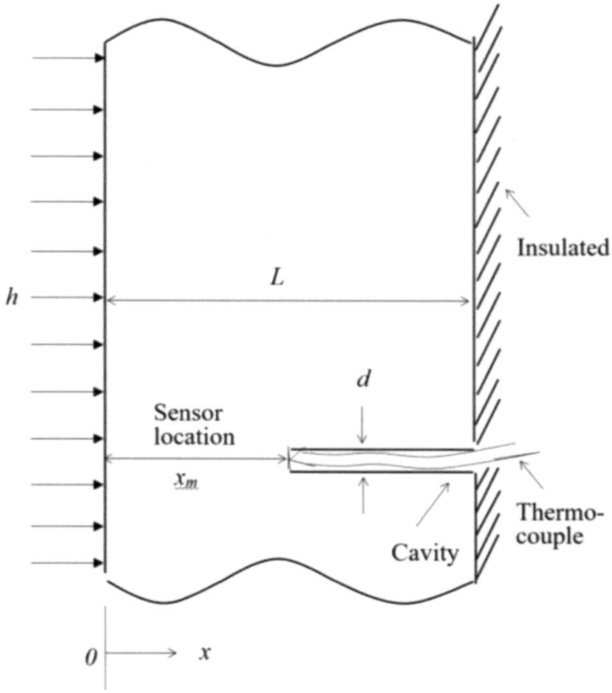

Figure 1.
Geometry of the specimen.

flux history has been analyzed with simulated data in a one-dimensional domain by Woodbury [26].

Chen and Danh [27] have carried out experimental studies to obtain transient temperature distortion and thermal delay in a slab due to presence of thermocouple cavity. The distortion of temperature profiles inside the slab may be influenced by the dissimilar thermophysical properties of thermocouple and surrounding materials and by the diameter and depth of the cavity. The temperature distortion [28] inside a slab is a function of the thermocouple cavity diameter d and location x_m.

Standard statistical analysis consists of error in the measurement as an additive of true plus random, in zero mean, in constant variance, uncorrelated, normal, bell shaped probability density function, constant variance known, errors in the dependent variables and no-prior information about the parameters. The error in measurement can be obtain using exact analytical solution [29] as

$$E = \cos(x/L)(\cos\varepsilon - 1)e^{-\delta\tau} \quad (1)$$

$$E = \sin(x/L)(\cos\varepsilon - 1)e^{-\delta\tau} \quad (2)$$

where ε and $\delta\tau$ refer to error in measurement of thermocouple location and in time recording, respectively. One of the important points that must be mentioned, here, is the use of a starting solution. In the case of a solid rocket motor where boundary conditions are suddenly imposed by a wall, there will be high intensity of heat flux on cold wall, and the heat flux during the first few steps in time may not be very accurate. The numerical solution is initiated using an exact analytical solution instead of starting from the initial constant condition. Such solutions can be obtained from to exact analytical solution [30] of transient heat conduction equation. Heat transfer rates to the calorimetric probe are estimated from measurements of temperature and rate of temperature change using energy conservation considerations [31].

An optimization method based on a direct and systematic search region reduction optimization method [32] can be employed to estimate the unknown convective heat transfer coefficient in a typical rocket nozzle. The most attractive feature of the direct search scheme is the simplicity of computer programming. The pseudo-random algorithm, an effective tool for optimization, does not require computation of derivatives but depends only on function evaluation. It works even when the differentiability requirements cannot be ensured in the feasible domain. For initiating the search only an estimate of the feasible domain is needed. Therefore, another advantage of the method is that the starting condition is not crucial; any reasonable value will do.

3. Heat conduction equation

3.1 Analytical solution

The computation of the turbulent convective heat transfer coefficient from combustion gases to the rocket nozzle wall is based on the Bartz's equation [33] incorporating the effects of compressibility, throat curvature and variation of transport properties in the boundary layer. The transient heat conduction in a one-dimensional Cartesian coordinate system having two parallel plane surfaces S_n (n = 1, 2) of a slab may be written in dimensional form [34] as.

$$\frac{\partial\theta(X,\tau)}{\partial\tau} = \frac{\partial}{\partial X}\left[K(\theta)\frac{\partial\theta(X,\tau)}{\partial X}\right], \text{in region } R, \tau>0 \tag{3}$$

with following initial and boundary conditions:

$$\theta(X,0) = f_i(X), \text{in region } R, \tag{4}$$

$$-\frac{\partial\theta(S_n,\tau)}{\partial X} = Bi_n[\theta(S_n,\tau)], \text{or } \frac{q_w L}{(T_g - T_i)} \text{ on boundary } S_n, \tau>0 \tag{5}$$

where f_i is initial temperature distribution in the region R of the slab. Eq. (4) represents both convective heat transfer or heat flux condition as applied to the inner surface.

We now consider the constant thermal property solution and can be written in terms of eigen function $\psi(\lambda_m, X)$ as

$$\theta(X,\tau) = \sum_{m=1}^{\infty} \exp\left[-\lambda_m^2\tau\right]\psi(\lambda_m, X)\int_R \psi(\lambda_m\overline{X}) f(\overline{X}) d\overline{X} \tag{6}$$

In the above Eq. (5), Bi or q_w is the unknown parameter to be determined using measured temperature time history at location x_m as depicted in **Figure 1**. In estimating the unknown condition, one has to minimize the absolute difference between the calculated and measured temperature at specified location and time (x_m, τ) in a prescribed tolerance value using an iteration procedure. The iteration scheme is described in the following sections.

3.2 Inverse algorithm

The IHCP is solved by comparing calculated and measured temperature using an iterative technique [30]. In estimating q_w, one minimizes

$$F(q_w) \approx |\theta_c(X_m, \tau) - \theta_m(X_m, \tau)| \tag{7}$$

where θ_c and θ_m are the calculated and measured temperatures at (X_m, τ), respectively. The computed temperature is a nonlinear function of unknown parameters such as wall heat flux or convective heat transfer coefficient. Temperature is calculated using Eq. (6) and compared with the measured temperature as expressed in Eq. (7). The inverse problem starts with initial guess value of the unknown parameter. The second step is to correct the previous guessed unknown parameter using the Newton-Raphson method. The sensitivity coefficient can be obtained by differentiating temperature with respect to wall heat flux q_w. The iteration procedure will continue until $|F(q_w)| \leq 10^{-4}$. This iterative scheme estimates the component of the q_w at a time and thus may be considered on-line method.

The inverse method for solving a value of $q_w(0, \tau)$ is as follows. Initiate with an initial guess value of q_w, satisfy the convergence criterion, and implement the Newton-Raphson to obtain the estimate value.

Now, it is possible to estimate convective heat transfer coefficient and combustion gas temperature in conjunction with measured temperature history [35]. The equation for converting the calculated heat flux to the heat transfer coefficient is

$$Bi = \frac{Lq_w}{K(\theta)(T_g - T_i)} \tag{8}$$

In the foregoing equation, T_g is an unknown quantity and can be estimated using again the above-mentioned minimization and iteration methods. The convergence criterion for the iterative scheme remains same as mentioned above.

3.3 Numerical methods

It is not always feasible to obtain analytical solution of temperature-dependent thermal conductivity and radiation boundary condition. The Crank-Nicolson finite difference method with two-time level implicit numerical scheme [36] has been employed to solve the nonlinear conduction problem with the Newton-Raphson method to consider the radiation boundary condition.

Deforming or moving finite elements method [37] is used to solve linear heat conduction equation. The moving finite element [38] is used to consider the time delay in the measurement of back wall temperature.

3.4 Two-nodes system of transient heat conduction equation

For only two nodes the system of [39] equations reduce to the following pair of equations:

$$\frac{d\theta_0}{d\tau} = \frac{1}{(\Delta X)^2}[-2(\theta_1 + Bi\Delta X)\theta_0 + 2\theta_1] \tag{9}$$

$$\frac{d\theta_1}{d\tau} = \frac{1}{(\Delta X)^2}[2\theta_0 - 2\theta_1] \tag{10}$$

where 0 and 1 represent node in a slab of finite thickness. These are the exact solutions to the system of two ordinary differential equations which resulted from a two-node finite-difference approximation to the original problem.

$$\theta_0 = \frac{1}{\lambda_2 - \lambda_1}\left[(\lambda_2 + 2Bi)e^{\lambda_1 \tau} - (\lambda_1 + 2Bi)e^{\lambda_2 \tau}\right] \tag{11}$$

$$\theta_1 = \frac{1}{\lambda_2 - \lambda_1}\left[\lambda_2 e^{\lambda_1 \tau} - \lambda_1 e^{\lambda_2 \tau}\right] \tag{12}$$

where $\lambda_1 = -2 - Bi + \sqrt{Bi^2 + 4}$ and $\lambda_2 = -2 - Bi - \sqrt{Bi^2 + 4}$.

Solution of the above simultaneous equation calculates the temperature with a given value of Bi. The solution is now solving simultaneously Eqs. (11) and (12) to determine the unknown parameter.

4. Inverse problem of heat conduction applied to a rocket nozzle

The influence of constant (average) thermal conductivity, temperature-dependent thermal conductivity, computational grid in numerical solver, nonlinear boundary condition, cylindrical coordinate and the estimation of the wall heat flux and convective heat transfer is carried out by employing measured temperature history of a rocket nozzle of a solid motor. Solution of linear heat conduction equation is used to estimate the convective heat transfer coefficient with the measured temperature data of outer wall of a rocket nozzle. The running time of rocket motor is 16 s. The nozzle wall thickness L = 0.0211 m. The thermo-physical properties of the material are: ρ = 7900 kg m^{-3}, C_p = 545 J kg^{-1} K^{-1}, K (average) = 35 Wm^{-1} K^{-1}. Initial temperature T_i = 300 K and combustion gas temperature T_g = 2946.2 K are used in the solution of the heat conduction equation.

4.1 Average thermal conductivity

Prediction of convective heat transfer coefficient is carried out in conjunction with the calculated and measured temperature history at outer surface of nozzle divergent in a solid rocket motor static test. The constant thermal conductivity solution of the linear transient heat conduction problem [30] is

$$\theta(X,\tau) = 1 - 2\sum_{n=1}^{\infty} \frac{Bi}{\left(Bi^2 + \lambda_n^1 + Bi\right)} \frac{\cos\left[\lambda_n(1-X)\right]}{\cos \lambda_n} e^{-\lambda_n^2 \tau} \tag{13}$$

$$\lambda \tan \lambda = Bi \tag{14}$$

For estimating unknown boundary condition, the heat conduction equation is and solved with the following boundary and initial conditions.

$$\frac{\partial \theta(0,\tau)}{\partial X} = Bi[\theta(0,\tau) - 1], \tau{>}0 \tag{15}$$

$$\frac{\partial \theta(1,\tau)}{\partial X} = 0, \tau{>}0 \tag{16}$$

and

$$\theta(X,0) = 0, \text{for all } X \tag{17}$$

Exact analytical solution of transient heat conduction as written in Eq. (13) is used to estimate convective heat transfer on the inner surface of the rocket nozzle.

An iterative scheme is used to solve inverse problem [30]. The iteration is carried out till the absolute difference between calculated and measured temperature is less than or equal to 10^{-4}. **Table 1** exhibits the comparison between the estimated values of the convective heat transfer coefficient based on the exact solution of heat conduction equation with the calculated values of Bartz [33]. Bartz's equation calculates conservative estimates for the convective heat transfer to the wall [40].

4.2 Temperature-dependent thermal conductivity

An iteration procedure [41] is employed in conjunction with exact solution to predict convective heat transfer coefficient from the measured temperature-time data at the outer wall of the nozzle as shown in **Table 2**. The expression for temperature-dependent conductivity is $K(T) = k_0 - \beta T$. The value of k_0 and β are 57 $Wm^{-1}K^{-1}$ and 2.718 $Wm^{-1}K^{-2}$, respectively. The advantage of using the exact solution is found directly at specified location and time as compared to the numerical method which needs the computation from the initial state.

4.3 Numerical solution with various computational grids

Deforming or moving finite element is used to consider the time delay in temperature at the outer wall of the slab [37]. Estimated values of wall heat flux and heat transfer coefficient are tabulated in **Table 3**. It can be observed from the table that the estimated wall quantities are having significant influence on the predicted unknown boundary conditions. This example is extended to consider spatial grid changed and temporal dependence on the numerical solution using moving finite element method [38].

4.4 Nonlinear boundary condition

Numerical analysis of nonlinear heat conduction with a radiation boundary condition [36] is carried out to estimate wall heat flux using temperature history on the back wall of the rocket nozzle. The high temperature variation alters thermophysical properties of the material of mild steel. **Table 4** shows comparison

t, s	θ_0 at inner surface	θ_c at outer surface	θ_m at outer surface	h, W/m^2K	h_B, W/m^2K
6	0.2950	0.0098	0.0096	1821.9	2254.2
7	0.3109	0.0159	0.0158	1810.0	2254.2
8	0.2996	0.0212	0.0211	1610.3	2254.2
9	0.3244	0.0301	0.0302	1690.9	2254.2
10	0.3340	0.0386	0.0385	1669.7	2254.2
11	0.3416	0.0473	0.0472	1641.9	2254.2
12	0.3302	0.0529	0.0529	1497.6	2254.2
13	0.3312	0.0602	0.0604	1443.1	2254.2
14	0.3409	0.0677	0.0676	1387.0	2254.2
15	0.3442	0.0781	0.0782	1413.0	2254.2
16	0.3475	0.0862	0.0861	1383.7	2254.2

Table 1.
Solution of inverse heat conduction problem.

t, s	$\theta(0,\tau)$				*h*, W/m²K	
	Iterative method	Beck method	$\theta_c(1,\tau)$	$\theta_m(1,\tau)$	Iterative method	Beck method
6	0.0883	0.0838	0.0099	0.0098	536.6	581.7
7	0.1067	0.1075	0.0158	0.0159	600.6	587.0
8	0.1144	0.1116	0.0220	0.0212	592.6	598.4
9	0.1367	0.1367	0.0302	0.0302	674.2	685.3
10	0.1522	0.1545	0.0386	0.0385	712.9	693.2
11	0.1654	0.1690	0.0472	0.0472	737.4	730.0
12	0.1686	0.1639	0.0529	0.0529	718.2	721.9
13	0.1773	0.1777	0.0605	0.0605	723.6	725.8
14	0.1844	0.1813	0.0677	0.0676	723.0	725.1
15	0.1944	0.2040	0.0781	0.0782	753.6	765.0
16	0.2083	0.2174	0.0862	0.0862	758.3	770.0

Table 2.
Comparison between iterative and Beck methods.

t, s	T_m K at $X = 1$	Uniform grid		Non-uniform grid		Moving grid	
		$q_w \times 10^6$, W/m²	h_c, W/m²K	$q_w \times 10^6$, W/m²	h_c, W/m²K	$q_w \times 10^6$, W/m²	h_c, W/m²K
6	326	3.715	1964.5	3.846	2044.9	4.517	2412.1
7	342	2.700	1408.8	2.848	1449.6	2.818	1485.9
8	356	2.698	1436.9	2.840	1531.6	2.820	1512.8
9	380	2.704	1463.0	2.589	1569.8	2.842	1552.8
10	402	2.705	1491.4	2.858	1603.3	2.846	1586.7
11	425	2.704	1518.9	2.852	1632.6	2.845	1618.5
12	440	2.691	1539.7	2.805	1636.2	2.812	1630.8
13	460	2.683	1564.6	2.776	1649.7	2.791	1650.6
14	479	2.673	1588.1	2.738	1657.1	2.764	1665.9
15	507	2.094	1226.4	2.015	1190.6	2.091	1235.4
16	528	2.086	1231.8	1.981	1178.5	2.067	1231.6

Table 3.
Wall heat flux at various grid arrangements.

between the estimated convective heat transfer coefficients with the Bartz solution [33]. Effects of nonlinear IHCP with radiation boundary condition are investigated and results are presented in **Table 4**.

4.5 Heat conduction in a hollow cylinder

A grid point shift strategy [42] is adapted to solve inverse conduction problem in a radial coordinate of rocket nozzle with inner and outer radius of rocket nozzle. The inner and outer radius of the nozzle is 0.0839 m and 0.0105 m, respectively. The purpose of the present example to investigate the influence of radial coordinate

t, s	T_o, K at $X = 0$	T_m K at $X = 1$	$q_c \times 10^6$ W/m^2	h W/m^{2}K	h_B W/m^{2}K	T_{gc} K	T_g K
6	659.8	326	2.3547	950.0	2254.2	3137	2946.2
7	801.0	342	2.3899	1019.6	2254.2	3122	2946.2
8	900.7	356	2.2211	992.4	2254.2	3115	2946.2
9	996.3	380	2.6489	1237.1	2254.2	3113	2946.2
10	1050.5	402	2.3670	1135.5	2254.2	3108	2946.2
11	1066.4	425	1.7100	827.3	2254.2	3104	2946.2
12	1201.8	440	2.8144	1459.2	2254.2	3099	2946.2
13	1320.0	460	2.6559	1467.0	2254.2	3098	2946.2
14	1354.8	479	1.7595	991.7	2254.2	3095	2946.2
15	1383.4	507	1.3810	791.4	2254.2	3094	2946.2
16	1414.9	528	1.1684	681.8	2254.2	3094	2946.2

Table 4.
Solution with nonlinear boundary condition.

t, s	T_o K at $X = 0$	T_m K at $X = 1$	$q_c \times 10^6$ W/m^2	h, W/m^{2}K	h_B, W/m^{2}K	θ_g, K	θ_{gc}, K
6	1260.2	326	3.6805	1789.6	2254.2	3316	2946
7	1175.9	342	3.3995	1628.0	2254.2	3264	2946
8	1160.7	356	2.4745	1181.4	2254.2	3255	2946
9	1165.8	380	2.5385	1194.7	2254.2	3290	2946
10	1196.0	402	2.5348	1261.1	2254.2	3206	2946
11	1192.3	425	2.3385	1166.4	2254.2	3197	2946
12	1205.8	440	2.2094	1114.8	2254.2	3187	2946
13	1211.0	460	2.1333	1229.5	2254.2	2946	2946
14	1222.1	479	2.0441	1187.5	2254.2	3943	2946
15	1237.1	507	2.0626	1206.7	2254.2	2946	2946
16	1249.1	528	2.0027	1180.9	2254.2	2945	2946

Table 5.
Inverse problem in a hollow cylinder.

on the estimated values of heat transfer coefficient. **Table 5** shows the effect of geometrical parameters on the predicted heat transfer coefficient.

4.6 Estimation of heat flux and heat transfer coefficient

The calculated convective heat transfer coefficients and inner wall temperature are used to determine the wall heat flux and the combustion temperature using Eq. (8). The iterative scheme is based on relation between wall heat flux and convective heat transfer coefficient [35]. **Table 6** shows the predicted values of wall heat flux and convective heat transfer coefficient. The IHCP is extended to determine wall heat flux in conjunction with convective heat transfer coefficient. A similar IHCP but referring to the 122 mm medium-range missile during correction engine operation has been considered by Zmywaczyk et al. [43].

t, s	T_0 K at $X = 0$	T_m K at $X = 1$	$q_c \times 10^6$, W/m^2	h, W/m^{2}K	h_B, W/m^{2}K	T_g, K	T_{gc}, K
6	1355.6	326	3.2502	2631.2	2254.2	3351	2946
7	1287.8	342	3.2950	1805.3	2254.2	3113	2946
8	1315.6	356	3.2974	1861.5	2254.2	3087	2946
9	1368.9	380	3.2967	1885.9	2254.2	3117	2946
10	1414.4	402	3.2837	1962.1	2254.2	3088	2946
11	1463.6	425	3.2718	2049.5	2254.2	3060	2946
12	1370.8	440	2.3825	1476.0	2254.2	2985	2946
13	1360.9	460	2.4140	1502.1	2254.2	2968	2946
14	1370.3	479	2.3625	1520.6	2254.2	2924	2946
15	1382.5	507	2.3675	1517.2	2254.2	2943	2946
16	1399.3	528	2.3645	1540.7	2254.2	2934	2946

Table 6.
Wall heat flux and convective heat transfer coefficient.

5. Estimation of heat flux with two-nodes in a sounding rocket

A two-node exact solution is used to calculate the back-wall temperature as described in Section 3.4. The iterative method described above has been used for estimating aerodynamic heating for a sounding rocket in free flight test. Here, the wall heat flux is estimated using the measured temperature history in conjunction with the iterative technique [30]. The aerodynamic heating rate is estimated for a typical sounding rocket as depicted in **Figure 2**. The location of thermocouple is marked in the diagram. The thermophysical properties of Inconel and wall thickness are $k = 18$ Wm^{-1} K^{-1}, $\alpha = 4.47 \times 10^{-6}$ m^2/s, $L = 0.7874 \times 10^{-3}$ m. **Figure 3** depicts the measured temperature time history at different locations measured from the tip of the cone in the free flight of a sounding rocket as delineated in **Figure 2**. It can be observed from temperature history that the initial time delay in thermal response is 6 s. The unknown q_w are estimated using an iterative technique which starts with an initial value of wall heat flux and is repeated until $|F(q_w)| \leq 10^{-4}$.

A two-node exact solution is used to calculate the wall temperature distribution. The unknown q_w are estimated using an iterative technique which starts with an initial value of wall heat flux and is repeated until $|F(q_w)| \leq 10^{-4}$. **Figure 4** displays the estimated variation of the wall heat flux as a function of flight time of the sounding rocket.

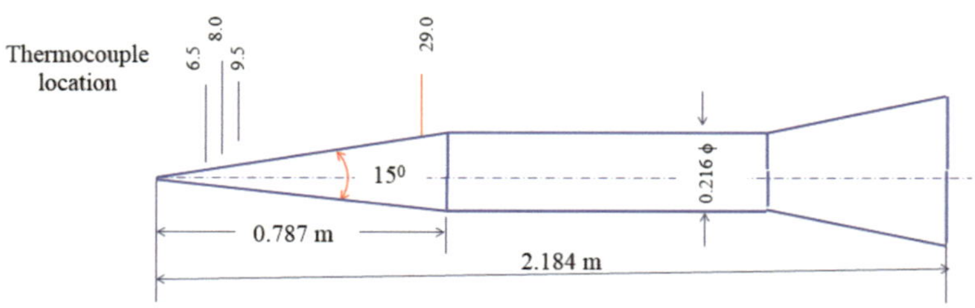

Figure 2.
Schematic sketch of sounding rocket showing location of thermocouple.

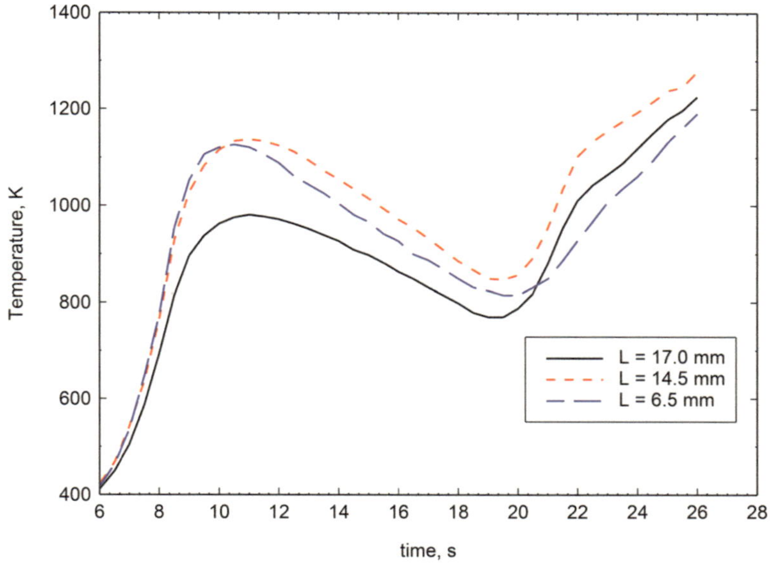

Figure 3.
Measured temperature history in free flight of the sounding rocket.

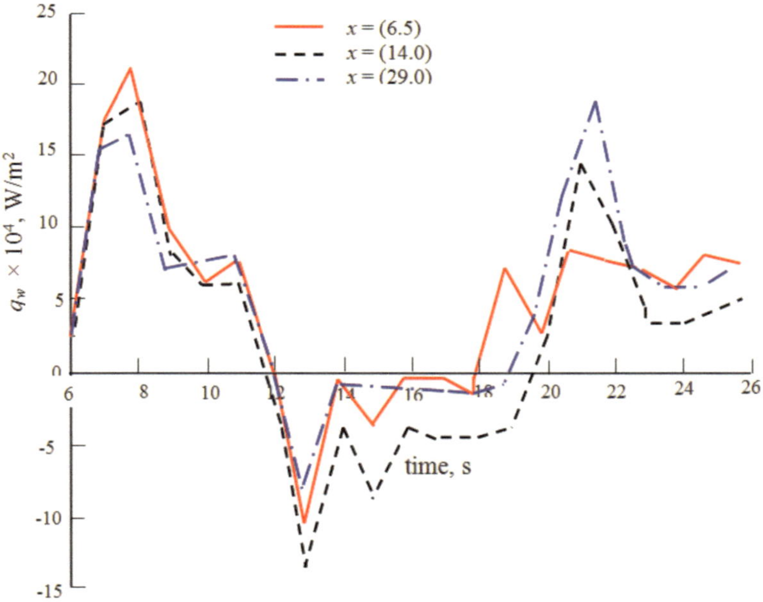

Figure 4.
Variations of wall heat flux vs. flight time.

The wall heat flux variation depends on the sounding rocket speed. The increase and decrease of the aerodynamic heating are a function of flight Mach number.

The estimated wall heat flux is compared with Van Driest's results [44]. **Table 7** depicts the estimated values of wall heat flux as a function of flight time at thermocouple location 29 as shown in **Figure 2**. It can be observed from the table that highest aerodynamic heating occurs during 7–8 s, another significant peak wall heat flux was found at 22 s.

t, s	T_m, K	T_c, K	$q_w \times 10^4$, W/m^2	$q_w^{van} \times 10^4$, W/m^2
6	313.0	320.0	2.756	8.246
7	341.3	349.4	16.695	14.82
8	408.0	412.3	19.387	22.647
9	469.7	469.3	8.657	18.876
10	495.2	495.8	8.145	10.117
11	504.1	513.3	9.302	4.569
12	502.4	521.2	2.757	1.176
13	495.2	506.8	−7.695	−0.977
14	487.4	487.6	−0.040	−2.320
15	477.4	477.0	−0.037	−3.134
16	467.4	467.0	−0.028	−3.578
17	457.4	457.5	−0,034	−3.769
18	445.8	447.2	−0.905	−3.789
19	438.6	437.5	−0,005	−1.628
20	444.1	445.9	5.173	4.135
21	469.1	467.5	13.518	11.524
22	533.0	521.2	20.618	19.732
23	556.3	558.4	9.180	13.122
24	567.1	573.5	6.514	8.581
25	584.1	587.2	8.592	5.467
26	596.9	602.2	7.018	3.314

Table 7.
Comparison between calculated and Van Driest's heat flux at location 29.

6. Conclusions

Analytical, transient numerical and two-node methods are used to compute temperature distribution in a finite slab. Numerical solution is carried out with temperature-dependent thermal conductivity. Implicit finite difference scheme with two-time level technique is implemented to solve nonlinear problem of heat conduction. Time delay is studied using finite element method with deforming grid strategy. A boundary shifting numerical scheme is used to solve transient heat conduction in radial coordinate. Evidence of temporal accuracy and dependence on time-step is demonstrated in the numerical solving of IHCP. Influence of thermocouple cavity and measurement errors in location and time are discussed. The IHCP is applied to predict the wall heat flux in a rocket nozzle of a solid motor. Wall heat flux is estimated in a free flight of a sounding rocket using the two-node method.

Nomenclature

Bi	Biot number, hL/k
C_p	specific heat
h	heat transfer coefficient
$K(\theta)$	thermal conductivity, $k(\theta)/k_0$

k_0	reference thermal conductivity at T_i
L	slab thickness
p	nondimensional parameter, $\alpha\Delta T/(\Delta x)^2$
q_w	wall heat flux
T	temperature
t	time
x	distance from the inner surface
X	dimensionless coordinate
α	thermal diffusivity
ρ	density
θ	nondimensional temperature, = $(T - T_g)/(T_g - T_i)$
τ	nondimensional time, $\alpha t/L^2$

Subscripts

B	Bartz
c	computed
g	combustion gas temperature
i	initial value
m	measured
o	outer wall
w	wall
β	constant thermal conductivity coefficient

Author details

Rakhab C. Mehta
Department of Aeronautical Engineering, Noorul Islam Centre for Higher Education, India

*Address all correspondence to: drrakhab.mehta@gmail.com

References

[1] Stolz G. Numerical solutions to an inverse problem of heat conduction for simple shapes. Journal of Heat Transfer. 1960;**82**(1):20-26

[2] Burggraf OR. An exact solution of the inverse problem in heat conduction theory and applications. Journal of Heat Transfer. 1964;**86**(2):373-382

[3] Beck JV. Nonlinear estimation applied to the nonlinear heat conduction problem. International Journal of Heat and Mass Transfer. 1970;**13**:703-716

[4] Frank I. An application of least squares method to the solution of the inverse problems of heat conduction. Journal of Heat Transfer. 1963;**85**: 378-379

[5] Imber M, Kan J. Prediction of transient temperature distributions with embedded thermocouples. AIAA Journal. 1972;**10**:784-789

[6] Shumakov NV. A method for the experimental study of the process heating a solid body. Journal of Technical Physics of the Academic of Sciences USSR. 1957;**2**:771-777

[7] Vogel J, Sára L, Krejčí L. A simple inverse heat conduction method with optimization. International Journal of Heat and Mass Transfer. 1993;**36**(17): 4215-4220

[8] Beck JV, Blackwell B, St. Clair CR Jr. Inverse Heat Conduction, Ill-Posed Problems. New York: Wiley-Interscience Publication; 1977

[9] Özisik MN, Orlande HRB. Inverse Heat Transfer: Fundamentals and Applications. New York: Taylor and Francis; 2000

[10] Tikhonov AN, Arsenin VY. Solution of Ill-Posed Problems. Washington, DC: Winston & Sons; 1977

[11] Kurpisz K, Nowak AJ. Inverse Thermal Problems. Southampton, UK: Computational Methods Publications; 1995

[12] Hensel E. Inverse Theory and Applications for Engineers. New Jersey: Prentice Hall; 1991

[13] Alifanov OM. Inverse Heat Transfer Problems. Berlin, Germany: Springer-Verlag; 1994

[14] Taler J, Duda P. Solving Direct and Inverse Heat Conduction Problems. Berlin: Springer; 2006

[15] Rumsey CB, Lee DB. Measurements of aerodynamic heat transfer on a 15-degree cone-cylinder, flare configuration in free-flight at Mach numbers up to 4.7. NASA TN D-824; 1961

[16] Mehta RC. Estimation of aerodynamic heat transfer in free-flight at Mach number upto 4.7. Wärm-und Stüffubertrang. 1986;**20**:27-31

[17] Mehta RC, Jayachandran T, Sastri VMK. Finite element analysis of conduction and radiative heating of a thin skin calorimeter. Wärm-und Stüffubertrang. 1988;**22**:227-230

[18] Dąbrowski A, Dąbrowski L. Inverse heat transfer problem solution of sounding rocket using moving window optimization. PLoS One. 2019;**14**(6):1-24

[19] Howard FG. Single thermocouple method for determining heat flux to a thermally thick wall. NASA TN D 4737; 1968

[20] Zmywaczyk J, Koniorczyk P. Numerical solution of inverse radiative-conductive transient heat transfer problem in a grey participating medium. International Journal of Thermophysics. 2009;**30**:1438-1451

[21] Karnal M. High temperature measurements at the internal nozzle wall of the zephyr [MS thesis]. Germany: Department of Computer Science, Electrical and Space Engineering, Lulea University of Technology; 2014

[22] Heisler MP. Temperature charts for induction and constant temperature heating. Journal of Heat Transfer, Transactions of the ASME. 1947;**69**: 227-236

[23] Mehta RC, Jayachandran T. Determination of heat transfer coefficient using transient temperature response chart. Wärm-und Stüffubertrang. 1990;**26**:1-5

[24] Li DI, Wells MA. Effects of subsurface thermocouple installation on the discrepancy of the measured thermal history and predicted surface heat flux during a quench operation. Metallurgical and Materials Transactions B. 2005;**36**(3):343-354

[25] Perakis N, Haidin OJ. Inverse heat transfer method applied to capacitively cooled rocket thrust chambers. International Journal of Heat and Mass Transfer. 2018;**131**:150-166

[26] Woodbury KA. Effect of thermocouple sensor dynamics on surface heat flux prediction obtained via inverse heat conduction analysis. International Journal of Heat and Mass Transfer. 1990;**33**(12):2641-2649

[27] Chen CJ, Danh TM. Transient temperature distortion in a slab to thermocouple cavity. AIAA Journal. 1978;**14**(7):979-981

[28] Anil SL, Mehta RC. Effect of thermocouple cavities on heat transfer measurements. In: Proceedings of XVII National and VI ISHMT/ASME Heat and Mass Transfer Conference, HMT-2004-C68, IGCAR, Kalpakam, India. Jan. 2004

[29] Carslaw HD, Jaeger JC. Conduction of Heat in Solids. London, UK: Oxford University Press; 1959

[30] Mehta RC. Solution of the inverse conduction problem. AIAA Journal. 1977;**15**(9):1355-1358

[31] Mehta RC. Estimation of heating rate using a calorimeter probe. Review of Scientific Instrumentation. 1987; **52**(11):1782-1784

[32] Mehta RC, Tiwari SB. Controlled random search technique for estimation of convective heat transfer coefficient. Heat and Mass Transfer Journal. 2007; **43**:1171-1177

[33] Bartz DR. A simple equation for rapid estimation of rocket nozzle convective heat transfer coefficients. Jet Propulsion. 1957;**27**:49-51

[34] Özisik MN. Boundary Value Problem of Heat Conduction. USA: International Text Book Co; 1968

[35] Mehta RC. Estimation of heat transfer coefficient in a rocket nozzle. AIAA Journal. 1981;**19**(8): 1085-1086

[36] Mehta RC. Numerical solution of nonlinear inverse heat conduction problem with radiation boundary conditions. International Journal of Numerical Methods in Engineering. 1984;**20**:1057-1066

[37] Mehta RC, Jayachandran T. Deforming finite element for the numerical solution of the nonlinear inverse heat conduction problem. Communication in Applied Numerical Methods. 1987;**3**:167-172

[38] Mehta RC, Jayachandran T. Deforming grid method applied to the inverse problem of heat conduction. Journal of Thermophysics and Heat Transfer. 1989;**3**(2):226-229

[39] Myers GF. Analytical Method in Conduction Heat Transfer. New York, USA: McGraw-Hill; 1971

[40] Brinsmade AF, Desmon LG. Hypothesis for correlating rocket nozzle throat convective heat transfer. Heat Transfer-Cleveland, Chemical Engineering Progress Symposium. 1965; **61**(59):88-98

[41] Mehta RC. Extension of the solution of inverse conduction problem. International Journal of Heat and Mass Transfer. 1979;**22**:1149-1150

[42] Mehta RC. An efficient numerical method for solving inverse conduction problem in a hollow cylinder. AIAA Journal. 1984;**22**(6):860-862

[43] Zmywaczyk J, Koniorczyk P, Preiskorn M, Machowski B. An inverse approach to estimate heat transfer coefficients of 122 mm medium-range missile during correction engine operation, problems of mechatronics armament, aviation. Safety Engineering. 2014;**5**(15):25-40

[44] Van Driest ER. In: Lin CC, editor. Turbulent Flows and Heat Transfer, 6. Princeton, NJ, USA: Princeton University Press; 1959

Chapter 10

Introduction to Numerical Approaches for Forward and Inverse Heat Transfer Problems

Sergey Voronin

Abstract

We describe the formulation of forward and inverse problems in heat transfer and illustrate several techniques applicable to their solution.

Keywords: heat transfer, finite differences, regularization, Fourier transforms, interpolation

1. Introduction

Heat flow is described by the heat (or diffusion) partial differential equation, which takes the form $\frac{\partial u}{\partial t} = k\frac{\partial^2 u}{\partial x^2}$ with $u = u(x,t)$ representing the temperature at time t and at spatial location x (which may have one or more components). The constant k is the ratio of the thermal conductivity in the material over the specific heat times the density of the material. Applying the conservation law principle to the heat flow across the rod yields the parabolic partial differential equation. If a heat source is present or the material is exothermic, a non-homogeneous version of the problem: $\frac{\partial u}{\partial t} - k\frac{\partial^2 u}{\partial x^2} = f(x,t)$ with $f(x,t) \neq 0$ applies. The initial boundary value problem comes from assigning relevant initial and boundary conditions. For example, for a perfectly insulated rod as in **Figure 1**, the conditions could be $u(0,t) = u(L,t) = 0$. Given some initial profile of the temperature at time $t = 0$, $u(x,0) = h(x)$, the analytical solution in 1D or 2D with a regular geometrical domain is given by separation of variables. Assuming $u(x,t) = F(x)G(t)$, one obtains a system of two ODEs and from that, the series solution $u(x,t) = \sum_{n=1}^{\infty} u_n(x,t) = \sum_{n=1}^{\infty} A_n \exp\left(-\lambda_n^2 t\right) \sin\left(\frac{n\pi}{L}x\right)$ with the coefficients A_n satisfying the initial condition $u(x,0) = \sum_{n=1}^{\infty} A_n \sin\left(\frac{n\pi}{L}x\right) = f(x)$ and the eigenvalues $\lambda_n = \frac{nk\pi}{L}$. The coefficients are evaluated via the sin series integration: $A_n = \frac{2}{L}\int_0^L f(x) \sin\left(\frac{k\pi x}{L}\right) dx$. For non-intregrable cases or more complicated geometries, numerical techniques for approximation are used [1].

1.1 Derivation of homogeneous and non-homogeneous equations

We now discuss two simple scenarios which can be used to derive homogeneous and non-homogeneous forms of the heat Equation [2]. Heat is the flow of energy from a warmer to a cooler location. There are several types of heat transfer including conduction (flow of heat through stationary material), convection (flow of heat through fluids), and radiation (flow of heat through electromagnetic waves) [2].

IntechOpen

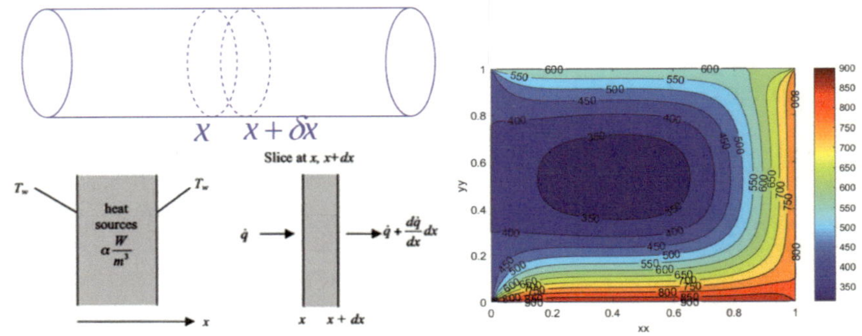

Figure 1.
Perfectly insulated rod used a model of heat transfer, an exothermic slab, and sample modeled heat profile in 2D.

In this chapter we are concerned mainly with heat conduction for which the differential heat model applies. We take Q to represent the heat transfer rate in units of Watts. It is the rate at which heat is flowing through one of the slabs in **Figure 2**. When heat flows into and out of a mass, the temperature of the mass changes in accordance with the model: $Q_{\text{in}} - Q_{\text{out}} = mc\frac{dT}{dt}$, where T represents the temperature, m the mass, and c is the specific heat value of the material. It should be noted that in heat problems, T and u are ofter used interchangeably. We define $q = Q/A$ (heat over area) as the heat flux, which by Fourier's law is proportional to the temperature gradient: $q = -\overline{k}\frac{dT}{dx}$, where $\overline{k}$ is the thermal conductivity of the material [2]. Using $Q_{\text{in}} = Q(x)$ and $Q_{\text{out}} = Q(x + \delta x)$ along with heat flux $q = Q/A$, we have $Aq(x) - Aq(x + \delta x) = mc\partial T/\partial t$. Applying Fourier's law yields:

$$\overline{k}A\left[-\frac{\partial T}{\partial x}\bigg|_x + \frac{\partial T}{\partial x}\bigg|_{x+\delta x}\right] = mc\partial T/\partial t$$

Next, using Taylor's approximation $\frac{\partial T}{\partial x}\Big|_{x+\delta x} \approx \frac{\partial T}{\partial x}\Big|_x + \frac{\partial^2 T}{\partial x^2}\Big|_x \delta x$ and expressing $m = \rho A \delta x$ in terms of slab density, area, and width, we obtain:

$$\overline{k}A\frac{\partial^2 T}{\partial x^2} = \rho c\frac{\partial T}{\partial t} \quad \Rightarrow \quad k\frac{\partial^2 T}{\partial x^2} = \frac{\partial T}{\partial t}$$

For the exothermic slab case exhibited in **Figure 2** on the right, we obtain an inhomogeneous equation. We assume the slab is filled with a material which emits heat at a rate $\dot{q}$ with units W/m^3 (energy per unit volume). In this case, the in and out energy balance gives $Q(x) + \dot{q}V - Q(x + \delta x) = mc\partial T/\partial t$, which with $q = Q/A$, $V = A\delta x$, Fourier's law and Taylor's expansion, gives $\overline{k}A\frac{\partial^2 T}{\partial x^2}\delta x + \dot{q}A\delta x = mc\frac{\partial T}{\partial t} \quad \Rightarrow \quad \overline{k}\frac{\partial^2 T}{\partial x^2} + \dot{q} = \rho c\frac{\partial T}{\partial t} \quad \Rightarrow \quad k\frac{\partial^2 T}{\partial x^2} + \frac{\dot{q}}{\rho c} = \frac{\partial T}{\partial t}$. In the case of steady state, $\partial T/\partial t = 0$ and a second order differential equation model results.

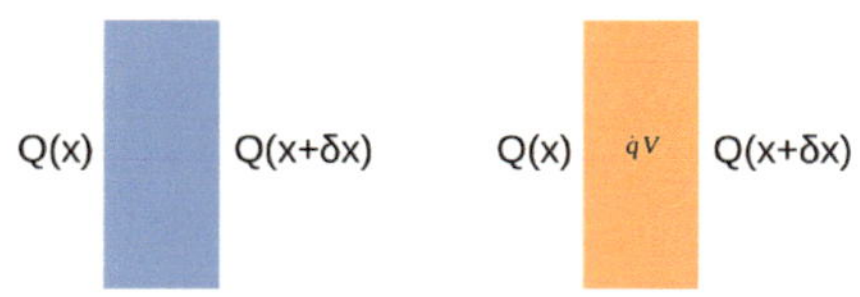

Figure 2.
Heat conducting slabs without and with internal heat generation.

1.2 Numerical schemes

Discretization of the derivative operators yields the finite difference formulation. A number of different schemes are possible based on the difference schemes used to replace the derivatives in the equation resulting in either an explicit method (where the update for u at the next time step is given by a direct calculation from the previous one) or an implicit method (where the update for u is accomplished via a linear solve). For 2D heat transfer, the model becomes $\alpha^2\left(\frac{\partial^2 u}{\partial x^2}+\frac{\partial^2 u}{\partial y^2}\right)=\frac{\partial u}{\partial t}$ and again, the finite difference method can be used for discretization. The initial value problems are either *forward* or *inverse*, in the sense that they either predict the temperature profile at later time from the knowledge of the initial temperature distribution at time $t=0$ or in the inverse case, predict the temperature distribution at time $t=0$ from knowledge of the temperature at a future time. Sometimes, the initial boundary value problem may be prescribed extra conditions, such as Neumann or other derivate condition type and the inverse problem could instead seek the initial values for such expressions from future data [3]. In general, the inverse problem is more challening than the forward problem due to being ill-posed. The discretized operator matrix which can be assembled from the difference formulation is not well conditioned and as a result the problem is overly sensitive to slight changes in input values. Regularization techniques need to be used for the solution [4].

2. Forward problem

The forward problem is concerned by the prediction of heat profile in future time from the initial condition. We first consider the heat equation initial boundary value problem in 1D:

$$u_t = k u_{xx}, \quad 0<x<1, \tag{1}$$

$$u(x,0) = f(x), \quad 0<x<1, \tag{2}$$

$$u(0,t) = u(1,t) = 0, \quad t>0 \tag{3}$$

with constant $k\in\mathbb{R}$. Via a transformation, this problem can be extended to nonzero boundary conditions. For (3), separation of variables gives the solution as the superposition of solutions of the form:

$$\tilde{u}(x,t) = e^{-\pi^2 kt}\sin(\pi x)$$

A quick calculation shows that $\left[\frac{\partial}{\partial t}-k\frac{\partial^2}{\partial x^2}\right]\tilde{u}(x,t)=0$. In order to form the well-known explicit scheme, we use the forward in time and central in space difference operators:

$$\frac{\partial u}{\partial t}_{\text{fd}}(x_i,t_j) = \frac{u(x_i,t_{j+1})-u(x_i,t_j)}{l} \tag{4}$$

$$\frac{\partial^2 u}{\partial x^2}_{\text{fd}}(x_i,t_j) = \frac{u(x_{i+1},t_j)-2u(x_i,t_j)+u(x_{i-1},t_j)}{h^2} \tag{5}$$

which results in the explicit finite difference scheme:

$$u_{i,j+1} = u_{i,j} + \frac{kl}{h^2}\left(u_{i+1,j}-2u_{i,j}+u_{i-1,j}\right) \tag{6}$$

This can be re-written as $u_{i,j+1} = (1 - 2\alpha)u_{i,j} + \alpha(u_{i+1,j} + u_{i-1,j})$ with $\alpha = \frac{kl}{h^2}$, with l the time step (Δt) and h the spatial step size (Δx). Von Neumann stability analysis yields the condition $l \le \frac{h^2}{2k}$. For non-homogeneous cases the general model is $\frac{\partial u}{\partial t} = k\frac{\partial^2 u}{\partial x^2} + f(x,t)$ and the finite difference analysis proceeds in a similar way, with an explicit scheme of the form: $u_{i,j+1} = u_{i,j} + \frac{kl}{l}h^2(u_{i+1,j} - 2u_{i,j} + u_{i-1,j}) + lf(x_i, t_j)$.

If we use the backward time finite difference approximation for $\frac{\partial u}{\partial t}$, we get: $\frac{u_{i,j+1} - u_{i,j}}{k} = \frac{u_{i+1,j+1} - 2u_{i,j+1} + u_{i-1,j+1}}{h^2}$, yielding the implicit scheme: $-\alpha u_{i-1,j+1} + (1 + 2\alpha)u_{i,j+1} - \alpha u_{i+1,j+1} = u_{i,j}$ which can be solved for time $j + 1$ from the information at time j using a linear system with a tridiagonal matrix with elements $(-\alpha, (1 + 2\alpha), -\alpha)$. The implicit scheme is numerically stable for any α but is costlier than the explicit scheme. Another example, is the so called Crank–Nicolson scheme based on the time derivative approximation at the midpoint of the time stamp. It takes the form:

$$\frac{u_{i,j+1} - u_{i,j}}{l} = k\frac{(u_{i+1,j+1} - 2u_{i,j+1} + u_{i-1,j+1}) + (u_{i+1,j} - 2u_{i,j} + u_{i-1,j})}{2h^2}$$

Again using $\alpha = \frac{kl}{h^2}$, we arrive at the scheme:

$$-\alpha u_{i+1,j+1} + 2(1 + \alpha)u_{i,j+1} - \alpha u_{i-1,j+1} = \alpha u_{i+1,j} + 2(1 - \alpha)u_{i,j} + \alpha u_{i-1,j}$$

Note that this takes the form of the linear system $Au = b$, where A is a tridiagonal matrix of terms $(-\alpha, 2(1 + \alpha), -\alpha)$, as before, u containing the heat distribution at time $j + 1$ and b containing the heat distribution information at time j. A linear system solves then yields the new time information from previous data.

In 2D, the standard explicit scheme arises from the discretization:

$$\frac{T_{i,j}^{n+1} - T_{i,j}^{n}}{\Delta t} = k\left(\frac{T_{i,j+1}^{n} - 2T_{i,j}^{n} + T_{i,j-1}^{n}}{(\Delta x)^2} + \frac{T_{i+1,j}^{n} - 2T_{i,j}^{n} + T_{i-1,j}^{n}}{(\Delta y)^2}\right)$$

Like in the 1D scheme, suitable finite difference approximations yield the implicit method.

A practically useful direction to consider is the use of non-standard finite difference (nsfd) schemes [5]. Based on (5), we define the discretized operators:

$$\overline{d}_t f(x,t) = f(x, t + l) - f(x,t) \tag{7}$$

$$\overline{d}_x^2 f(x,t) = f(x + h, t) - 2f(x,t) + f(x - h, t) \tag{8}$$

and proceed to evaluate $\left(\overline{d}_t - kl\frac{\overline{d}_x^2}{h^2}\right)\tilde{u}(x,t)$ for $\tilde{u}(x,t) = e^{-\pi^2 kt}\sin(\pi x)$. The standard fd scheme does not set this expression to zero, and hence this analytically valid solution does not satisfy the original partial differential equation. However, based on the calculations below, we can make a replacement for $\frac{kl}{h^2}$, which would force the equation to zero.

$$\begin{aligned}\overline{d}_t[\tilde{u}(x,t)] &= \sin(\pi x)\left[e^{-\pi^2 k(t+l)} - e^{-\pi^2 kt}\right] = \sin(\pi x)e^{-\pi^2 kt}\left(e^{-\pi^2 kl} - 1\right)\\ &= \tilde{u}(x,t)\left(e^{-\pi^2 kl} - 1\right)\end{aligned} \tag{9}$$

$$\overline{d}_x^2[\tilde{u}(x,t)] = e^{-\pi^2 kt}[\sin(\pi(x+h)) - 2\sin(\pi x) + \sin(\pi(x-h))] \tag{10}$$

$$= e^{-\pi^2 kt}\left[\frac{1}{2i}\left(e^{i\pi(x+h)} - e^{-i\pi(x+h)}\right) - 2\sin(\pi x) + \frac{1}{2i}\left(e^{i\pi(x-h)} - e^{-i\pi(x-h)}\right)\right] \tag{11}$$

$$= e^{-\pi^2 kt}\frac{1}{2i}\left[e^{i\pi x}e^{i\pi h} - e^{-i\pi x}e^{-i\pi h} - 4i\sin(\pi x) + e^{i\pi x}e^{-i\pi h} - e^{-i\pi x}e^{i\pi h}\right] \tag{12}$$

$$= e^{-\pi^2 kt}\frac{1}{2i}\left[e^{i\pi x}\left(e^{i\pi h} + e^{-i\pi h}\right) - e^{-i\pi x}\left(e^{-i\pi h} + e^{i\pi h}\right) - 4i\sin(\pi x)\right] \tag{13}$$

$$= e^{-\pi^2 kt}\frac{1}{2i}\left[2\cos(\pi h)\left(e^{i\pi x} - e^{-i\pi x}\right) - 4i\sin(\pi x)\right] \tag{14}$$

$$= e^{-\pi^2 kt}\frac{1}{2i}[4i\cos(\pi h)\sin(\pi x) - 4i\sin(\pi x)] \tag{15}$$

$$= 2e^{-\pi^2 kt}\sin(\pi x)[\cos(\pi h) - 1] = 2\tilde{u}(x,t)[\cos(\pi h) - 1] \tag{16}$$

Setting $\left(\overline{d}_t - \frac{kl}{h^2}\overline{d}^2{}_x\right)\tilde{u}(x,t) = 0$, we obtain:

$$\tilde{u}(x,t)\left[e^{-\pi^2 kl} - 1\right] - \frac{2kl}{h^2}\tilde{u}(x,t)[\cos(\pi h) - 1] = 0 \quad \Rightarrow \quad \left[e^{-\pi^2 kl} - 1\right] = \frac{2kl}{h^2}[\cos(\pi h) - 1]$$

Hence, the nsfd scheme follows if we make the following replacement in (6):

$$\frac{kl}{h^2} = \frac{1}{2}\left[\frac{e^{-\pi^2 kl} - 1}{\cos(\pi h) - 1}\right]$$

When $h, l \ll 1$, the nsfd substitution coincides with the standard finite difference scheme. We use the small $|q|$ Taylor series approximations $e^q \approx 1 + q$ and $\sin(q) \approx q$ to obtain:

$$\frac{1}{2}\left[\frac{e^{-\pi^2 kl} - 1}{\cos(\pi h) - 1}\right] = \left[\frac{e^{-\pi^2 kl} - 1}{-2\sin^2\left(\frac{\pi h}{2}\right)}\right] \approx \frac{1}{2}\left[\frac{1 - \pi^2 kl - 1}{-\frac{\pi h^2}{2}}\right] = \frac{kl}{h^2}$$

By this it follows that for small h, l, the stability constraint $\frac{kl}{h^2} \leq 1$ is satisfied in the nsfd scheme when $l \leq \frac{h^2}{2}$ just like in the original scheme. This also holds for large h, as long as h is chosen so that $|\cos(\pi h) - 1|$ is not very small.

2.1 Numerical experiments

Below, we present some sample results for the case of constant $k \in \mathbb{R}$ in (3), where we found that the described nsfd scheme often outperforms the fd scheme at the same time step. In **Figure 3**, we illustrate the solutions with fd and nsfd methods with the initial condition, $u(x, 0) = \sin(\pi x) + x$ for problem 1, with boundary conditions $u(0, t) = 0$ and $u(1, t) = 1$ and with the initial condition $u(x, 0) = 12\sin(9\pi x) - 7\sin(4\pi x)$ and zero homogeneous boundary conditions for problem 2. For the second example, errors vs. step size are also compared. In general, we observe more accurate results with the use of the nsfd scheme.

An interesting case is that of heat transfer with an exothermic material, which can be modeled with a heat source distributed throughout [3]. Considering an

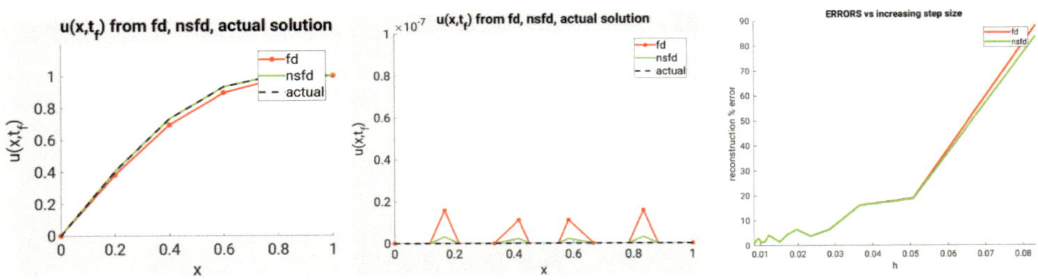

Figure 3.
Comparison of fd and nsfd scheme solutions for the 1-D heat equation for problems 1 and 2 above.

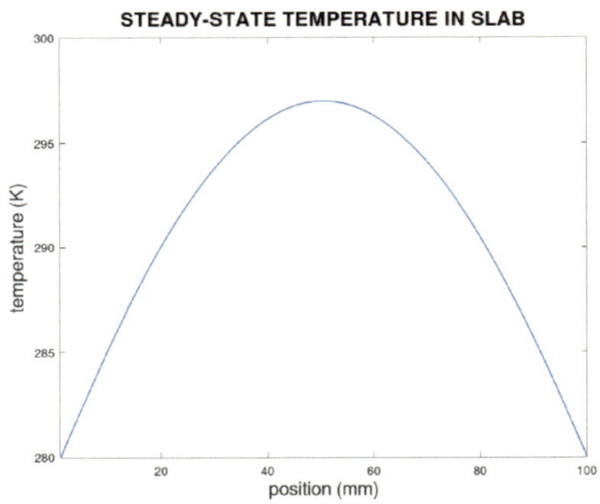

Figure 4.
Temperature distribution inside a thin slab with an exothermic material with homogeneous boundary conditions as simulated by an fd scheme.

infinitesimal segment as in **Figure 1** yields the energy balance equation: $\dot{q} + \alpha dx - \left(\bar{q} + \frac{d\dot{q}}{dx}dx\right) = 0 \Rightarrow \frac{d\dot{q}}{dx} = \alpha$. It follows that the heat model takes the form $\frac{1}{\alpha}\frac{\partial u}{\partial t} = \frac{\partial^2 u}{\partial x^2} + \frac{\dot{q}}{k}$. The steady state behavior does not depend on time so that $\frac{\partial u}{\partial t} = 0$ and an ordinary differential equation model $\frac{d^2 u}{dx^2} + \frac{\alpha}{k} = 0$ results. Upon a double integration, the general solution is $u(x) = -\frac{\alpha}{2k}x^2 + Ax + B$, which with the boundary conditions $T(0) = T(L) = T_w$ becomes $u(x) = -\frac{\alpha}{2k}x^2 + Lx + T_w$, a parabola. We can simulate this steady state example with a finite difference scheme for the time derivative, yielding the temperature profile shown in **Figure 4**.

3. Inverse problem

The inverse problem consists of predicting the initial condition from the observation of the heat profile at some later time t. In general, the problem is ill-posed and regularization is necessary for its solution [4]. Let us look at a simple example in 1-D. Starting from the initial boundary value problem on $x \in [0, L]$:

$$\frac{\partial u}{\partial t} = k\frac{\partial^2 u}{\partial x^2} \tag{17}$$

$$u(x,0) = f(x), \quad u(0,t) = 0 = u(L,t) \tag{18}$$

In this case, separation of variables can be used and after a well-known sequence of steps, the following series solution can be derived [3]:

$$u(x,t) = \sum_{n=1}^{\infty} A_n \sin\left(\frac{n\pi x}{L}\right) \exp\left(-k\left(\frac{n\pi}{L}\right)^2 t\right)$$

It follows that $u(x,0) = \sum_{n=1}^{\infty} A_n \sin\left(\frac{n\pi x}{L}\right)$ which is a sin series; that is, $A_n = F[u(x_0)] = \frac{2}{L}\int_0^L u(x,0)\sin\left(\frac{n\pi x}{L}\right)$. It follows that the sin coefficients of the solution at $t = T$ are given by the sin coefficients at $t = 0$ multiplied by $\exp\left(-k\left(\frac{n\pi}{L}\right)^2 T\right)$. The values of u at $t = T$ can then be computed via the inverse sin transform applied to the product. In a similar way, it follows that from the transform of the solution at $t = T$, $F[u(x,T)]$, we can in principle go backwards, to obtain $u(x,0) = F^{-1}\left[F[u(x,T)].*\exp\left(k\left(\frac{n\pi}{L}\right)^2 T\right)\right]$ with the $.*$ representing component-wise operation [3]. However, with the negative sign inside the exponential gone, there is potential for blow up of any large magnitude terms. The issue can be addressed by performing filtering on the large coefficients, setting the largest magnitude portion to zero or some maximum value. The magnitude cut off value can be selected based on quartile statistics of the nonzero absolute values.

A more general approach for inversion is to utilize the inverse problem formulations of the matrix heat propagation formulation. For example, for the standard implicit or Crank-Nicholson scheme, we have the system $Au = b$, where A is a tridiagonal matrix of terms $(-\alpha, 2(1+\alpha), -\alpha)$. This means that $u^1 = A^{-1}b(u^0)$, $u^2 = A^{-1}b(u^1) = A^{-2}b(u^0) \quad \Rightarrow \quad b(u^0) = A^2u^2$. This then implies that the initial condition at time $t = 0$, u^0 can be recovered from the application of the inverse matrix raised to a power n_t corresponding to the time point. As another example, if we discretize $u_t = ku_{xx}$ as:

$$\frac{\partial u(t)}{\partial t} - \frac{k}{h^2}\left[u_{i+1}^{j+1}(t) - 2u_i^{j+1}(t) + u_{i-1}^{j+1}(t)\right] = 0 = \frac{u_i^{j+1} - u_i^j}{\Delta t} - Mu^{j+1}(t),$$

with $M = \frac{k}{h^2}\text{tridiag}(-1, 2, -1)$, for $j = 0, 1, \dots, n_t - 1$. Using $u(x,0) = U_0$, we obtain the relations:

$$u^0 = U_0; u^{j+1} = (I + \Delta t M)^{-1}u^j = (I + \Delta t M)^{-(j+1)}U_0 \Rightarrow \overline{L}^{-(j+1)}U_0 = u^{j+1}$$

This means that U_0 can in theory be obtained right from $(I + \Delta t M)^{n_t}u^{n_t}$. The matrix needs not be formed explicitly, as the finite differencing scheme can be repeatedly applied to $u^{n_t} = u(T)$. However, as shown in **Figure 5**, raising the typical differencing matrix to a power results in non-linear singular value decay. Hence, the application of the inverse of matrix $(I + \Delta t M)^{-n_t}$ to the approximant for $b = u(T)$, the heat profile at time T, would result in the blow-up of any numerical errors due to the ill-conditioning. We must instead consider the inverse formulation $LU_0 = b$ with $L = \overline{L}^{-n_t}$ and utilize a regularization scheme.

This can be accomplished by replacing the naive solution $L^{-1}b$ by the regularized formulation $\min_u\left(\|Lu - b\|_2^2 + \lambda\|u\|_2^2\right)$, which for large enough $\lambda > 0$, alleviates the problem due to the fast decay of singular values. This well-known Tikhonov regularization formulation essentially acts to remove the influence of small singular values on the solution, which would otherwise blow up in value upon inversion.

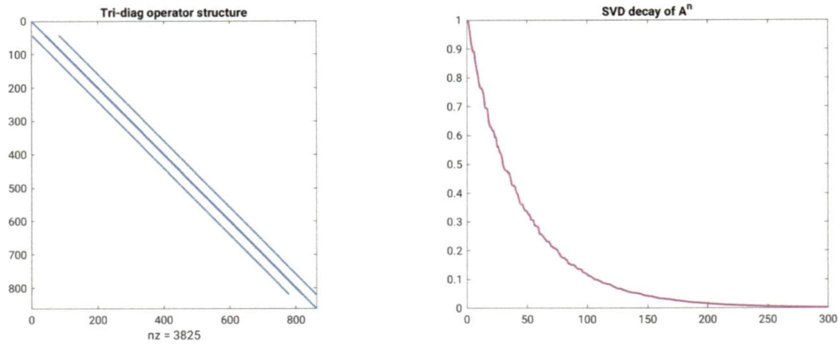

Figure 5.
Tridiagonal structure of differencing operator matrix and singular value decay of operator raised to a power.

The regularization parameter λ should be chosen with care to provide regularization but not to move too far away from the naive inverse operator. For this reason, λ is usually selected based on an L-curve criterion [6], whereby the parameter is initially set at some fraction of $||L^T b||_\infty$ and then iteratively lowered to a lower fraction (with the values either linearly or logarithmically spaced). At each λ value, the solution to $(L^T L + \lambda I)u_\lambda = L^t b$ is obtained (e.g. with a Conjugate Gradient scheme) using the solution at the previously used λ as the initial guess. Typically, the progression to lower λ stops once the value of $||Lu_\lambda - b||_2/N$ reaches a plateau. It is possible to employ a *predictor-corrector* scheme to iteratively improve the solution. In this approach, the approximation to U_0 can be obtained as above, then the forward operator can be applied to propagate this approximated initial heat profile at $t = 0$ to time $t = T$ and compare with observations [4]. We can compare the residual at time T and attempt some corrections to U_0 based on this residual such that the propagated solution at $t = T$ best matches observations. In the case of only a finite subset of observations, interpolation techniques can be utilized to complete the profile at $t = T$ across all values of x [7] prior to regularization to obtain an approximate smooth and continuous $u(T)$ heat profile.

3.1 Numerical experiments

As an example, consider the 1D cosine based initial heat profile with $k = 1$ and zero boundary conditions. The profile and subsequent integration results are shown in **Figure 6**. It is clear that the temperature in the rod will approach zero at all locations with increasing time. The inverse problem consists of approximately recovering the initial profile $u(x, 0) = 0.5\cos(2\pi/65x) + 1)$, with support on

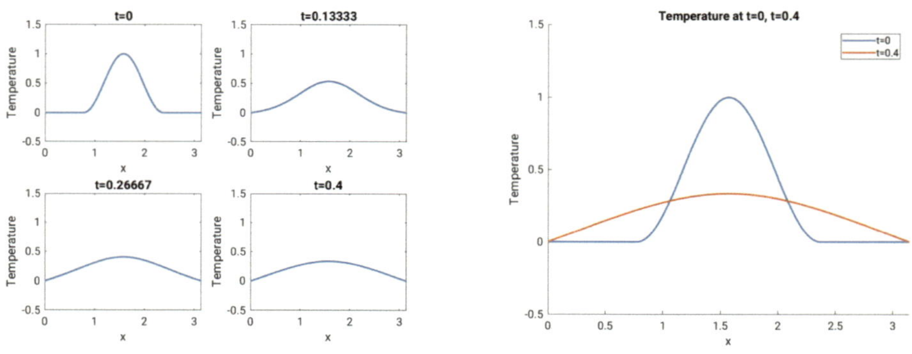

Figure 6.
Homogeneous 1D profile and change of temperature with time.

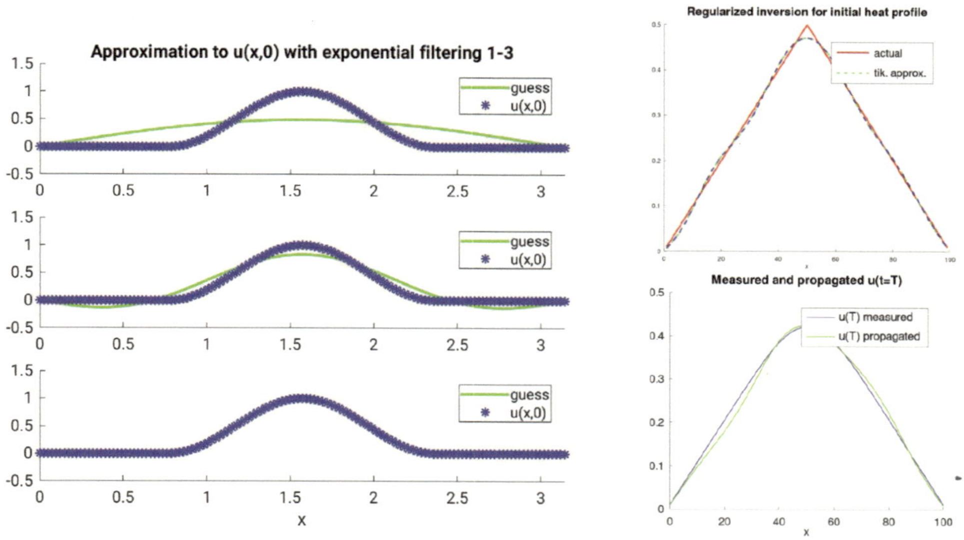

Figure 7.
Initial condition reconstruction with different filtering levels and illustration of inversion for initial condition using Tikhonov regularization with parameter tuned via the L-curve approach.

$[N/4, 3N/4]$. For this, we take the sin transform of the solution at $t = T$ and multiply by $\exp\left(k\left(\frac{n\pi}{L}\right)^2 T\right)$. In Octave, this can be accomplished with the code:

```
fuT=fft (u (:, n)); % temperature profile at time T
R=exp (k.*k '*T);
filter (R);
init_cond_approx=real (ifft (fuT.*R));
```

The term R may contain large entries, which will blow up when the inverse fft is performed. For this reason, the filtering of small entries is necessary. This can be accomplished simply by replacing all large entries of R with either zero or some max value; e.g. taking the cutoff to be all values above $1e6$ and replacing them either with 0 or $1e2$. The cutoff can be estimated with standard statistical outlier techniques, by taking the quartiles Q_1, Q_2, Q_3 of the $|R|$ values, computing $IQR = Q_3 - Q_1$ and filtering all entries of R above $Q_3 + \alpha IQR$ with $\alpha \geq 2$ by setting them to the Q_2 value. The process is illustrated in **Figure 7**, for the same example as above with filtering for absolute value of entries above $1e1, 1e3, 1e5$. It can be observed that filtering only the entries of R above $1e5$ in absolute value leads to a good approximation of the initial condition. Tikhonov regularization can be utilized when it is possible to build up the finite difference propagation operator (L), either as an explicit matrix, or as a function. An example is shown to the right of **Figure 7**, with an initial triangular temperature profile that is approximately recovered by a simple continuation scheme. In the same Figure, we also show the profiles of $u(T)$ both as observed and as propagated from the approximated initial values. The smoothing properties of the heat kernel [3] give rise to the observed transformation between $t = 0$ and $t = T$.

4. Conclusions

This article discusses numerical approaches for the heat equation, both for the forward and inverse problems with examples for homogeneous and non-homogeneous cases. For the forward heat propagation modeling problem, we have presented a simple non-standard finite difference formulation, which satisfies the differential model and appears to result in improved performance over the standard finite difference scheme with similar magnitude time and spatial step sizes. For the inverse problem, we have described approaches based on Fourier filtering and Tikhonov regularization.

Author details

Sergey Voronin
Research Scientist, Amathsolutions, USA

*Address all correspondence to: svoronin@amathsolutions.org

References

[1] Özişik, M. Necati, Helcio RB Orlande, Marcelo J. Colaço, and Renato M. Cotta. Finite difference methods in heat transfer. CRC press, 2017.

[2] Glicksman, Leon R., and John H. Lienhard. Modeling and Approximation in Heat Transfer. Cambridge University Press, 2016.

[3] Taler, Jan, and Piotr Duda. Solving direct and inverse heat conduction problems. Springer Science & Business Media, 2010.

[4] Muniz, Wagner Barbosa, Haroldo F. de Campos Velho, and Fernando Manuel Ramos. "A comparison of some inverse methods for estimating the initial condition of the heat equation." Journal of Computational and Applied Mathematics 103, no. 1 (1999): 145–163.

[5] Mickens, Ronald E. "Nonstandard finite difference schemes for differential equations." Journal of Difference Equations and Applications 8, no. 9 (2002): 823–847.

[6] Hansen, Per Christian, and Dianne Prost O'Leary. "The use of the L-curve in the regularization of discrete ill-posed problems." SIAM journal on scientific computing 14, no. 6 (1993): 1487–1503.

[7] Castro, L. P., Q. Chen, and S. Saitoh. "Source inversion of heat conduction from a finite number of observation data." Applicable Analysis 89, no. 6 (2010): 801–813.